Ceramic Materials for Energy Applications

Ceramic Materials for Energy Applications

A Collection of Papers Presented at the 35th International Conference on Advanced Ceramics and Composites
January 23–28, 2011
Daytona Beach, Florida

Edited by
Yutai Katoh
Kevin M. Fox
Hua-Tay Lin
Ilias Belharouak

Volume Editors
Sujanto Widjaja
Dileep Singh

A John Wiley & Sons, Inc., Publication

Published by John Wiley & Sons, Inc., Hoboken, New Jersey.
Published simultaneously in Canada.

Library of Congress Cataloging-in-Publication Data is available.

ISBN 978-1-118-05994-4

oBook ISBN: 978-1-118-09538-6
ePDF ISBN: 978-1-118-17241-4

ISSN: 0196-6219

Printed in the United States of America.

10 9 8 7 6 5 4 3 2 1

Contents

FUEL CERAMICS AND IRRADIATION EFFECTS

JOINING AND INTEGRATION OF CERAMIC STRUCTURES

PROCESSING

CERAMICS FOR ELECTRIC ENERGY GENERATION, STORAGE, AND DISTRIBUTION

ADVANCED MATERIALS AND TECHNOLOGIES FOR RECHARGEABLE BATTERIES

Preface

This proceedings contains contributions from three energy related symposia that were part of the 35th International Conference on Advanced Ceramics and Composites (ICACC), in Daytona Beach, FL, January 23–28, 2011. These symposia include Ceramics for Electric Energy Generation, Storage and distribution; Advanced Ceramics and Composites for Nuclear and Fusion Applications; and Advanced Materials and Technologies for Rechargeable Batteries. These symposia were sponsored by ACerS Engineering Ceramics Division and ACerS Nuclear & Environmental Technology Division.

The editors wish to thank the authors and presenters for their contributions, the symposium organizers for their time and labor, and all the reviewers for their valuable comments and suggestions. Acknowledgment is also due for the financial support from the Engineering Ceramic Division, the Nuclear & Environmental Technology Division and The American Ceramic Society.

YUTAI KATOH, *Oak Ridge National Laboratory*
KEVIN FOX, *Savannah River National Laboratory*
HUA-TAY LIN, *Oak Ridge National Laboratory*
ILIAS BELHAROUAK, *Argonne National Laboratory*

Introduction

This CESP issue represents papers that were submitted and approved for the proceedings of the 35th International Conference on Advanced Ceramics and Composites (ICACC), held January 23-28, 2011 in Daytona Beach, Florida. ICACC is the most prominent international meeting in the area of advanced structural, functional, and nanoscopic ceramics, composites, and other emerging ceramic materials and technologies. This prestigious conference has been organized by The American Ceramic Society's (ACerS) Engineering Ceramics Division (ECD) since 1977.

The conference was organized into the following symposia and focused sessions:

Symposium 1	Mechanical Behavior and Performance of Ceramics and Composites
Symposium 2	Advanced Ceramic Coatings for Structural, Environmental, and Functional Applications
Symposium 3	8th International Symposium on Solid Oxide Fuel Cells (SOFC): Materials, Science, and Technology
Symposium 4	Armor Ceramics
Symposium 5	Next Generation Bioceramics
Symposium 6	International Symposium on Ceramics for Electric Energy Generation, Storage, and Distribution
Symposium 7	5th International Symposium on Nanostructured Materials and Nanocomposites: Development and Applications
Symposium 8	5th International Symposium on Advanced Processing & Manufacturing Technologies (APMT) for Structural & Multifunctional Materials and Systems
Symposium 9	Porous Ceramics: Novel Developments and Applications

Symposium 10	Thermal Management Materials and Technologies
Symposium 11	Advanced Sensor Technology, Developments and Applications
Symposium 12	Materials for Extreme Environments: Ultrahigh Temperature Ceramics (UHTCs) and Nanolaminated Ternary Carbides and Nitrides (MAX Phases)
Symposium 13	Advanced Ceramics and Composites for Nuclear and Fusion Applications
Symposium 14	Advanced Materials and Technologies for Rechargeable Batteries
Focused Session 1	Geopolymers and other Inorganic Polymers
Focused Session 2	Computational Design, Modeling, Simulation and Characterization of Ceramics and Composites
Special Session	Pacific Rim Engineering Ceramics Summit

The conference proceedings are published into 9 issues of the 2011 Ceramic Engineering & Science Proceedings (CESP); Volume 32, Issues 2-10, 2011 as outlined below:

- Mechanical Properties and Performance of Engineering Ceramics and Composites VI, CESP Volume 32, Issue 2 (includes papers from Symposium 1)
- Advanced Ceramic Coatings and Materials for Extreme Environments, Volume 32, Issue 3 (includes papers from Symposia 2 and 12)
- Advances in Solid Oxide Fuel Cells VII, CESP Volume 32, Issue 4 (includes papers from Symposium 3)
- Advances in Ceramic Armor VII, CESP Volume 32, Issue 5 (includes papers from Symposium 4)
- Advances in Bioceramics and Porous Ceramics IV, CESP Volume 32, Issue 6 (includes papers from Symposia 5 and 9)
- Nanostructured Materials and Nanotechnology V, CESP Volume 32, Issue 7 (includes papers from Symposium 7)
- Advanced Processing and Manufacturing Technologies for Structural and Multifunctional Materials V, CESP Volume 32, Issue 8 (includes papers from Symposium 8)
- Ceramic Materials for Energy Applications, CESP Volume 32, Issue 9 (includes papers from Symposia 6, 13, and 14)
- Developments in Strategic Materials and Computational Design II, CESP Volume 32, Issue 10 (includes papers from Symposium 10 and 11 and from Focused Sessions 1, and 2)

The organization of the Daytona Beach meeting and the publication of these proceedings were possible thanks to the professional staff of ACerS and the tireless dedication of many ECD members. We would especially like to express our sincere thanks to the symposia organizers, session chairs, presenters and conference attendees, for their efforts and enthusiastic participation in the vibrant and cutting-edge conference.

ACerS and the ECD invite you to attend the 36th International Conference on Advanced Ceramics and Composites (http://www.ceramics.org/daytona2012) January 22–27, 2012 in Daytona Beach, Florida.

SUJANTO WIDJAJA AND DILEEP SINGH
Volume Editors
June 2011

Carbon Materials and Fuel Ceramics

IRRADIATION-INDUCED DIMENSIONAL CHANGE AND FRACTURE BEHAVIOR OF C/C COMPOSITES FOR VHTR APPLICATION

Junya Sumita[1], Taiju Shibata[1], Kazuhiro Sawa[1], Ichiro Fujita[2], Jun Ohashi[3], Kentaro Takizawa[4], W. Kim[5] and J. Park[5]

[1]Research Group for VHTR Fuel & Material, Division of Fuels and Materials Engineering, Nuclear Science and Engineering Directorate, Japan Atomic Energy Agency, 4002 Oarai-machi, Higashiibaraki-gun, Ibaraki-ken, 311-1393, Japan

[2]Atomic Energy Section, Production Division, Toyo Tanso Co., Ltd., 2791 Matsuzaki, Takuma-cho, Mitoyo-shi, Kagawa-ken, 769-1102, Japan

[3]Carbon Material Development Group, Ceramic Development Division, R&D Operation, Ibiden Co., LTd., 1-1 Kitagata, Ibigawa-cho, Ibi-gun, Gifu-ken, 501-0695, Japan

[4]Functional Materials Development Team, Research & Development Div., Tokai Carbon Co., Ltd., 394-1 Subashiri, Oyama-cho, Sunto-gun, Shizuoka-ken, 410-1431, Japan

[5]Nuclear Materials Research Division, Korea Atomic Energy Research Institute, 1045, Daedeok-daero, Yuseong-gu, Daejeon, 305-353, KOREA

ABSTRACT

For control rod element of Very High Temperature Reactor (VHTR), carbon fiber reinforced carbon composite (C/C composite) is one of the major candidate materials for its high strength and thermal stability. In this study, the irradiation-induced dimensional change of the C/C composite was measured to evaluate the dependence of the dimensional change on the heat-treatment temperature. Also, the fracture tests on three types of oxidized C/C composites were carried out to evaluate the change in stress intensity factor as a function of an amount of the thermal oxidation, burn-off. This study showed that the heat-treatment temperature effect on irradiation-induced dimensional change of the C/C composite was the similar as for graphite. Furthermore, from the view point of the irradiation-dimensional change, C/C composites that were exposed to heat-treatments at high temperatures are preferable. Moreover, the maximum load and stress intensity factor decreased with increasing burn-off. They were depending on the structure/micro-structure and on the adhesion strength between fiber and matrix. Although, the adhesion strength largely affected the maximum load and stress intensity factor at low burn-off, the influence of it was became low with increasing burn-off.

INTRODUCTION

The High Temperature Gas-Cooled Reactor (HTGR), graphite-moderated and helium gas-cooled, is an attractive nuclear reactor to obtain high temperature helium gas of about 950 °C at reactor outlet. The high temperature helium gas can be utilized not only for power generation but also for process heat source in a hydrogen production system [1-2]. Japan Atomic Energy Agency (JAEA) constructed the High Temperature Engineering Test Reactor (HTTR), the first HTGR in Japan [3]. It was demonstrated that the HTTR can produce high temperature helium gas of about 950

°C at the reactor outlet in the high temperature operation mode. A long-term high temperature operation test, continuous 50 days operation at an outlet temperature of 950 °C, was successfully carried out in 2010. A lot of valuable data on fuel performance, reactor physics, impurities in primary helium gas, etc. were obtained from the test.

The Very High Temperature Reactor (VHTR) is one of the promising candidate reactors for Generation-IV nuclear energy systems. International collaborations have been carried out through the Generation IV International Forum (GIF) [4]. As national projects on commercial HTGR/VHTR, USA focuses on the Next Generation Nuclear Plant (NGNP) and China has started the construction of High Temperature Reactor - Pebble-bed Module (HTR-PM). The Nuclear Hydrogen Development and Demonstration (NHDD) program has been initiated in Korea. JAEA has also carried out R&D on the design of an original VHTR based on the technologies established through the HTTR construction and operation. JAEA has designed a prismatic block type reactor, Gas Turbine High Temperature Reactor 300-Cogeneration (GTHTR-300C) [5-6]. It is expected to generate electricity as well as produce hydrogen by a thermochemical hydrogen production cycle, IS (Iodine-Sulfur) process.

Application of the more heat-resistant ceramic material to the control rod is one of the important subjects for the VHTR development. For the HTTR, ferritic superalloy Alloy 800H is used for the metal parts of the control rod [7]. Its maximum allowable temperature to be used repeatedly after scrams is set at 900 °C. For the VHTR, however, it would be essential to develop the more heat-resistant material as substitute for metallic materials, since the core components in the VHTR will be used in the severer condition comparing with the HTTR. The heat resistant ceramic composites, i.e. carbon fiber reinforced carbon composite (C/C composite) and SiC fiber reinforced SiC composite (SiC/SiC composite) are major candidate materials as substitute for the metallic materials of the control rod, which can be used at higher temperature.

JAEA has focused on a two-dimensional C/C composite (2D-C/C composite) which has the layer structure of laminas composed of fibers and matrix [8]. Fig. 1 shows a scheme of R&D for the C/C composite control rod. R&D are to be carried out in three phases: (1) Database establishment, (2) Design and fabrication and (3) Demonstration test by HTTR. For database establishment, material properties data, strength, elastic modulus, coefficient of thermal expansion (CTE), thermal conductivity, etc., in non-irradiated and irradiated conditions are accumulated. For design and fabrication, it is necessary to determine the control rod design. Connecting method for the component parts is one of the key technologies for the fabrication of control rod. Since appropriate design methodology for the fiber reinforced ceramics has not been fully developed yet, it is also important issue to establish the methodology for the C/C as well as the SiC/SiC composite application. Demonstration test in HTTR is the final phase for the composite control rod development. The large scale component of the C/C or SiC/SiC composite would be irradiated in the HTTR to evaluate its performance. This demonstration test is expected to be realized in an international collaboration.

Fig. 2 shows a concept of control rod elements. Control rod sheath is made of the C/C composite because of its good machinability, and connecting rod is made of the SiC/SiC composite because of high oxidation resistance and strength. The ceramic control rod would be used at temperatures of up to 1500 °C to a fluence of 1 to 2 dpa (displacements per atom). The ceramic control rod might be oxidized by small amount of impurities in He coolant. Maximum load on the connecting rod is estimated as 250 kg. Therefore, for the design of control rod for VHTR, it is necessary to characterize the mechanical/thermal properties to establish the database including oxidation and irradiation data.

In this study, the irradiation-induced dimensional change of the C/C composite is measured to evaluate the dependence of the dimensional change on the heat-treatment temperature. Also, the fracture tests on three types of oxidized C/C composites are carried out to evaluate the change in stress intensity factor as a function of an amount of the thermal oxidation, burn-off.

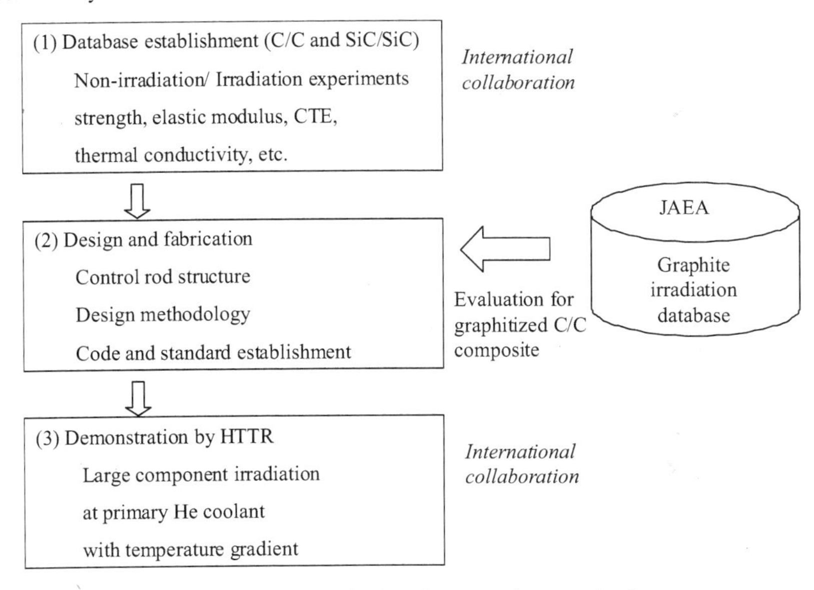

Fig. 1 Scheme of R&Ds for composite control rod.

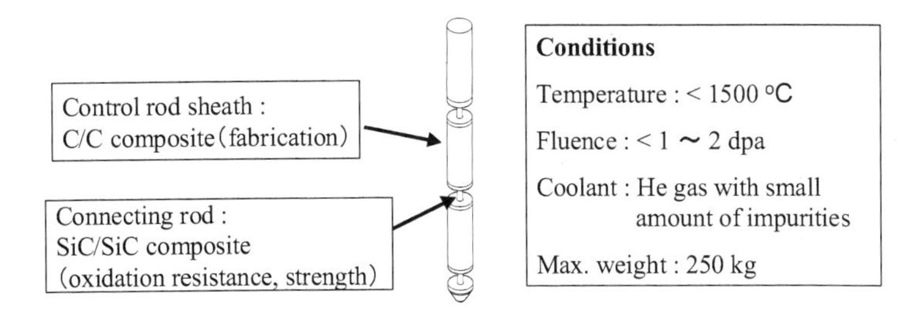

Fig. 2 Conceptual figure of the control rod element

EXPERIMENT

(1)Irradiation and post-irradiation test

A PAN-based, plain woven 2D-C/C composites, CX-270 and CX-270G grades were used in irradiation and post-irradiation tests because CX-270 and CX-270G were one of the major candidate materials for VHTR. A carbon fiber cloth based on 6 K (a fiber bundle consists of 6,000 fibers) plain-woven PAN (polyacrylonitrile)-based high strength type was coated by phenolic resin and formed to be prepreg. The prepreg was cut and molded to be preform before being baked, impregnated, then heated at 1800 to 2200 °C and purified to be CX-270. The fabrication process of CX-270G is almost identical except the final heat-treatment temperature which is in the range from 2800 to 3000 °C [9-10]. The volume fraction of the fiber in composites is about 50%. Table 1 shows typical material properties of CX-270 and CX-270G.

The samples were irradiated in a 03M-46AS capsule in Japan Material Testing Reactor (JMTR) of JAEA to a neutron fluence of 1.52×10^{24} n/m^2 (E > 1.0MeV) corresponding to 0.22 dpa at an irradiation temperature of 600 °C. After the irradiation, the irradiation-induced dimensional change of the cylindrical samples with 5 mm diameter and 20 mm length was measured.

Table 1 Typical properties of CX-270 and CX-270G [10]

Properties	Grade	
	CX-270	CX-270G
Bulk density(g/cm^3)	1.61	1.63
Tensile strength (MPa)	230(-p)	167(-p)
Bending strength (MPa)	177(-v)	133(-v)
Compressive strength (MPa)	261(-v)	89(-v)
Elastic modulus (GPa)	-	81(-p)
Thermal conductivity (W/m/K)	67(-p) 15(-v)	129(-p) 26(-v)
Coefficient of thermal expansion (RT～1000 °C , 10^{-6}/K)	<1(-p) 7.7(-v)	0.2(-p) 10.8(-v)
Final heat-treatment temperature (°C)	1800 ~ 2200	2800~3000

-p : direction parallel to lamina, -v : direction vertical to lamina

(2)Fracture test on oxidized C/C composites

Fracture tests were carried out on three grades of 2D-C/C composites, CX-270G, ICU-10 and CC28NHPS to investigate the influence of structure/micro-structure on fracture behavior as a function of burn-off. Table 2 shows typical material properties of CX-270G, ICU-10 and CC28NHPS. CC28NHPS. consists of 6K plain-woven PAN-based carbon fiber cloths as well as CX-270G. The fabrication process of CC28NHPS is similar to CX-270G. The final heat-treatment temperature of CC28NHPS was 2800 °C. The difference between CX-270G and CC28NHPS is bulk density as sown in Table 1. It means that the micro-structure between them is different.

Piled-up pitch-based chopped carbon fiber sheets were coated by resin, then molded, baked and impregnated. Finally they were heated at 2000 °C to be ICU-10. Samples were oxidized in air at 500-550 °C to attain homogeneous oxidation throughout the bulk. Burn-off, an amount of the thermal oxidation, is expressed by using the weight loss caused by oxidation. Since oxidation during the lifetime of control rod for VHTR is much low at normal operation, target burn-off values were 1 and 3wt %. Four types of crack generation were assumed to be occurred in the control rod of the C/C composite. In this study, the test results of a crack type shown in Fig.3 are reported. The test results of other crack types were already reported[11-12]. The four point bending test was carried out with the sample which size was of 10 (W) × 20 (B) ×45 (L) mm^3. A straight-through notch which depth (a_0) was of 5 mm with a maximum notch root radius of 0.02 mm was made by a cutter knife as shown in Fig. 4. Samples were pushed at a rate of 0.1mm/min at room temperature. The clip gauge with resolution of about 0.5 μm was installed at the bottom of the sample and the crack opening displacements (COD) were measured.

Table 2 Typical properties of C/C composite [10)]

Properties	Grade		
	CX-270G	ICU-10	CC28NHPS
Bulk density(g/cm^3)	1.63	1.62	1.76
Tensile strength (MPa)	167(-p)	100(-p)	173(-p)
Bending strength (MPa)	133(-v)	130(-v)	173(-v)
Elastic modulus (GPa)	81(-p)	-	112(-p)
Thermal conductivity (W/m/K)	129(-p) 26(-v)	130(-p) 35(-v)	127(-p) 21(-v)
Coefficient of thermal expansion (RT～1000 °C, 10^{-6}/K)	0.2(-p) 10.8(-v)	0.2(-p) 11(-v)	0.4(-p) 9.5(-v)
Final heat-treatment temperature (°C)	2800~3000	2000	2800

-p : direction parallel to lamina, -v : direction vertical to lamina

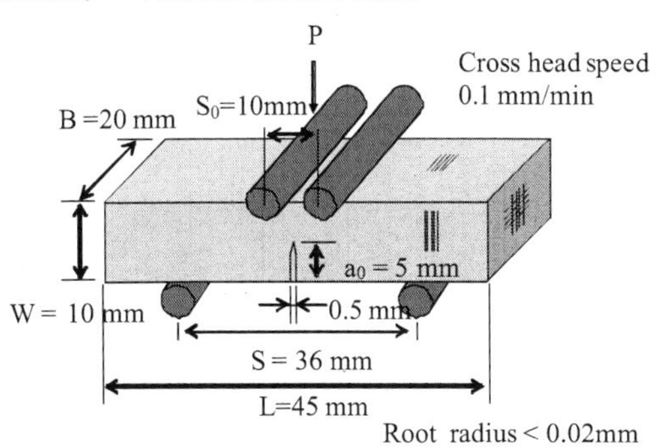

Fig. 3 Single edge-notched bend bar specimen of four point bending test

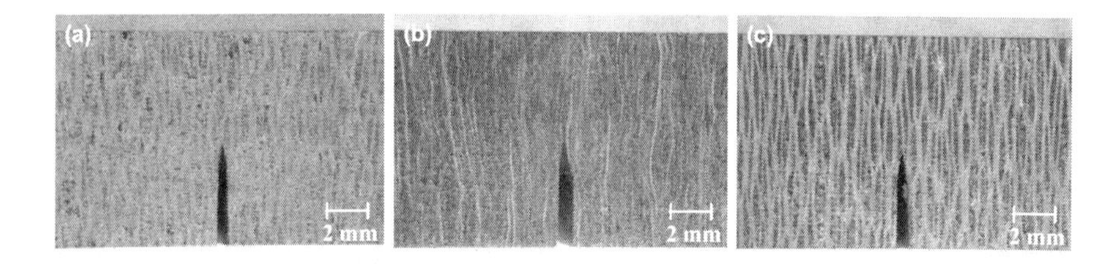

Fig. 4 Shape of notch (a) CX-270G, (b) ICU-10 and (c) CC28NHPS

RESULTS AND DISCUSSION

(1)Evaluation of irradiation behavior

The irradiation-induced dimensional change of CX-270 and CX-270G for the parallel to lamina direction (-p) is shown in Fig. 5. The dimensional change of CX-270G irradiated at 600 °C was expressed as the following equations in reference [8].

$$\Delta L / L_0 = aF \tag{1}$$

$$a = -4.4 \times 10^{-1} \text{ (\%/dpa), for parallel to lamina direction}$$

where, $\Delta L / L_0$ is the relative change of dimension (%), a is a constant, and F is fast neutron fluence (dpa).

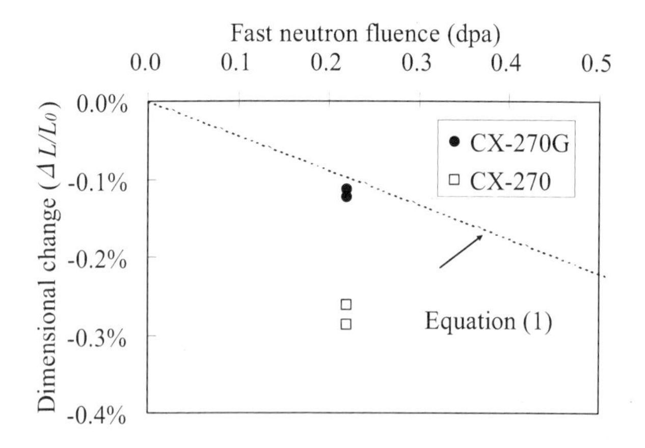

Fig. 5 Irradiation-induced dimensional change of CX-270G and CX-270

The relationship between irradiation-induced dimensional change and neutron fluence obtained in this study corresponded to the equation (1) as can be seen in Fig.5. Furthermore, the dimensional change of CX-270 was larger than CX-270G. As it is shown in literature for graphite, the magnitude of the dimensional change is depending on the final heat-treatment which is owed to the higher crystallinity, the reduction of initial lattice defects and the reduced damage accumulation rate [13]. Since the only difference between CX-270 and CX-270G is the heat-treatment temperature, a similar temperature dependent effect as in graphite is expected to occur for the C/C composite. Therefore, for the view point of the magnitude of the irradiation-induced dimensional change, the high heat-treated C/C composite is preferable.

(2)Evaluation of fracture behavior

The crack opening displacement (COD)-load behaviors of CX-270G, ICU-10 and CC28NHPS are shown in Fig. 6. The maximum load for ICU-10 showed the largest as was expected. ICU-10 is composed of short chopped carbon fibers which are distributed various direction. Since the crack in ICU-10 grows along with short fibers aligned in various directions, resistance to crack-extension is believed to become large. The difference in the maximum load between CX-270G and CC28NHPS would be depending on the interface adhesion between fiber and matrix. Since the adhesion strength between fiber and matrix for CX-270G would be higher than for CC28NHPS, the maximum load for CX-270G is high. Moreover, the COD for CC28NHPS (between 100 - 200 μm) is different from CX-270G and ICU-10 (-100 μm). This would be depending on the micro-structure of the C/C composite. The bulk density of CC28NHPS is higher than that of CX-270G and ICU-10 as shown in Table 2. It indicates that CX-270G and ICU-10 have more porosity in matrix. Since CC28NHPS has less porosity, the crack growth rate of CC28NHPS is low.

The maximum loads for CX-270G, ICU-10 and CC28NHPS decreased with increasing burn-off. Oxidation first occurs at interface between fiber and matrix [9]. Since the adhesion strength between fiber and matrix decreased with increasing burn-off, the maximum load became small. The drop in maximum load for ICU-10 is smaller than for CX-270G. It would be owed to the short fiber structure of ICU-10. CX-270G is stronger than CC28NHPS at low burn-off, but at 3% burn-off, the maximum load is almost the same. As mentioned before, the difference in the maximum load between CX-270G and CC28NHPS would be depending on the interface adhesion between fiber and matrix. However, with increasing burn-off, since the adhesion strength between fiber and matrix decreased by oxidation, the influence of it became low. Therefore, the maximum load for CX-270G and CC28NHPS was almost the same at 3% burn-off.

The equation to calculate the stress intensity factor for the metal was temporarily used because that for the C/C composite for VHTR has not established yet. The stress intensity factor (K_Q) was calculated by using the following equation [14].

$$
\begin{aligned}
&K_Q = \sigma a^{1/2} f(\frac{a}{W}) \\
&\sigma = 6M / BW^2 = 3P_Q(S - S_0) / 2BW^2 \\
&f(\frac{a}{W}) = 1.99 - 2.47(a/W) + 12.97(a/W)^2 - 23.17(a/W)^3 + 24.8(a/W)^4
\end{aligned} \tag{2}
$$

where P_Q is the load at the cross point of the line of 95% of slope at the origin and the original COD-load curve, M is applied bending moment, a is notch depth, S is outer support span, S_0 is inner minor span, B is the breadth of the specimen and W is specimen depth. Since some test conditions in this study were different from those described in ASTM STP 410, K_Q values obtained here were treated as tentative ones. The calculated K_Q are shown in Fig. 7. The K_Q of ICU-10 showed largest as was the case for the maximum load. This would be owed to the difference of structure. The K_Q decreased with increasing burn-off because the adhesion strength decreased by oxidation. Moreover, the K_Q of CX-270G was larger than CC28NHPS at low burn-off, but at 3% burn-off, the K_Q was almost the same. These would also be owed to the adhesion strength between fiber and matrix.

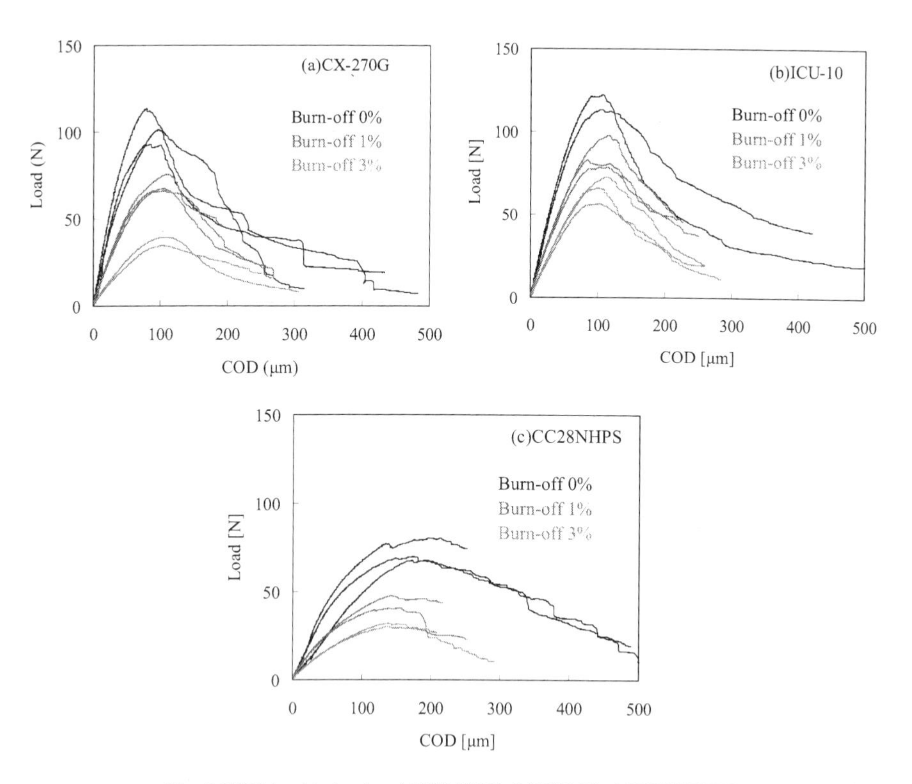

Fig. 6 COD-load behavior, (a)CX-270G, (b)ICU-10, (c)CC28NHPS

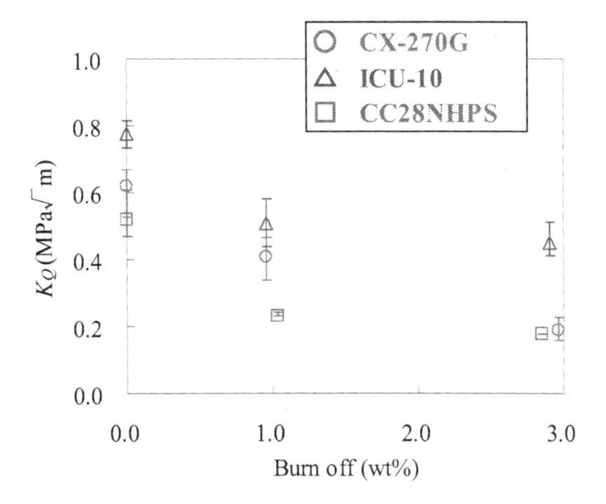

Fig. 7 Calculated K_Q as a function of burn-off

CONCLUDING REMARKS

In this study, the irradiation-induced dimensional change of the C/C composite was measured to evaluate the dependence of the dimensional change on the heat-treatment temperature. Also, the fracture tests on three types of oxidized C/C composites were carried out to evaluate the change in stress intensity factor as a function of burn-off. The following results were obtained.

- The heat-treatment temperature effect on irradiation-induced dimensional change of the C/C composite was the similar as for graphite. Furthermore, from the view point of the irradiation-dimensional change, C/C composites that were exposed to heat-treatments at high temperatures are preferable.
- The maximum load and stress intensity factor decreased with increasing burn-off. They were depending on the structure/micro-structure and on the adhesion strength between fiber and matrix. Although, the adhesion strength largely affected the maximum load and stress intensity factor at low burn-off, the influence of it was became low with increasing burn-off.

REFERENCES

[1] K. Matsui, S. Shiozawa, M. Ogawa and X. L. Yan, Present Status of HTGR Development in Japan, Proc. 16th Pacific Basin Nuclear Conference (16PBNC), Aomori, Japan, Oct. 13-18, 2008, PaperID P16P1355

[2] S. Shiozawa, et al., Status of the Japanese Nuclear Hydrogen Program, ANS 2007 Annual Meeting, Embedded Topical on Safety and Technology of Nuclear Hydrogen Production, Control and Management" (ST-NH2), June24-28, 2007, Boston, MA, USA.

[3] S. Saito, et al., Design of High Temperature Engineering Test Reactor (HTTR), JAERI 1332 (1994).

[4] T. J. O'Cornnor, Gas Reactors - A Review of the Past, an Overview of the Present and a View of the Future, Proc. of GIF Symposium, pp.77-92, 9-10 Sept. 2010, Paris, France.

[5]K. Kunitomi, S. Katanishi, S. Takada, et al., Reactor core design of Gas Turbine High Temperature Reactor 300, Nucl. Eng. Design, 230, 349-366 (2004).

[6]K. Kunitomi, X. Yan, T. Nishihara, et al., JAEA's VHTR for hydrogen and electricity cogeneration: GTHTR300C, Nucl. Eng. and Technol., 39[1], 9-20 (2007).

[7]Y. Tachibana, S. Shiozawa, J. Fukakur, F. Matsumoto and T. Araki, Integrity assessment of the high temperature engineering test reactor (HTTR) control rod at very high temperatures, Nucl. Eng. and Design 172, 93-102(1997).

[8]T. Shibata, J. Sumita, T. Makita, T. Takagi, E. Kunimoto, K. Sawa, Research and developments on C/C composite for Very High Temperature Reactor (VHTR) application, Ceramic Engineering and Science Proceedings (CESP), Vol.30, Issue 10, pp.19-32 (2009).

[9]T. Sogabe, M.Ishihara, S. Baba, Y. Tachibana, M. Yamaji, T. Iyoku and T. Hoshiya, Effect of air-oxidation on the thermal diffusivity of the nuclear grade 2-dimensional carbon fiber reinforced carbon/carbon composite, Material Science Research International, 9[3] 235 (2003).

[10]T. Sogabe, M. Ishihara, S. Baba, T. Kojima, Y. Tachibana, T. Iyoku, T. Hoshiya, T. Hiraoka and M. Yamaji, Development of the carbon fiber reinforced carbon-carbon composite for High Temperature Gas-cooled Reactor, JAERI-Research 2002-026 (2002).

[11]T. Shibata, J. Sumita, T. Makita, T. Takagi, E. Kumimoto, K. Sawa, W. Kim, C. Jung and J. Park, Transactions of the Japan Society mechanical Engineers Series A 76[764] 383 (2010).

[12]J.Sumita, I. Fujita, T. Shibata, T. Makita, T. Takagi, E. Kunimoto, K. Sawa, W. Kim and J. Park, Study on fracture behavior of 2D-C/C composite for application to control rod of Very High Temperature Reactor, Proc. ICC3 2010 14-18th, November 2010, Osaka, Japan.

[13]T. D. Burchell, "Carbon Materials for Advanced Technologies", PERGAMON (1999).

[14]W. F. Brown, Jr and J. Srawley, Plane strain crack toughness testing of high strength metallic materials, ASTM STP410, pp1-pp65 (1965).

ACKNOWLEDGEMENT

The authors would like to express gratitude to Dr. M. Eto, technical consultant to TOYO TANSO Co., LTD., for his valuable comments. A part of this project was supported by JSPS (Japan Society for the Promotion of Science) and KOSEF (Korea Science and Engineering Foundation) under the Japan-Korea Basic Scientific Cooperation Program.

R&D AND IRRADIATION PLANS FOR NEW NUCLEAR GRADE GRAPHITES FOR APPLICATION TO VHTR

Kentaro Takizawa, Kazuyuki Kakehashi, Toshiaki Fukuda, Tohru Kida, Tokai Carbon Co. LTD., Japan; Kazuhiro Sawa, Junya Sumita, Japan Atomic Energy Agency, Japan; Yutai Katoh, Lance L. Snead, Oak Ridge National Laboratory, USA

ABSTRACT

Fine-grained isotropic graphite shows high strength making it a promising material for the graphite component of High Temperature Gas-cooled Reactor (HTGR) and Very High Temperature Reactor (VHTR). Service life of the graphite component is determined primarily by the residual stress after neutron irradiation in the reactor core. It is expected that development of new fine-grained isotropic nuclear grade graphite possessing higher strength will contribute toward added design margins and an extension of the service life of components, which likely improve the reactor economy very significantly.

Tokai Carbon Co. LTD. has started the development of nuclear grade graphite having high strength for the graphite component of VHTR. As the start of evaluation to the application for VHTR, the evaluation based on Japanese draft standard for HTGR is advanced. Japanese draft standard demands as follows, (1)Guarantee a stable supply for long period, (2)Obtain the data for design, (3)Confirms stable features under high temperature and irradiation condition.

The following conclusions were obtained. (1)To satisfy the requirement of stable supply, G347S and G458S grades was selected. These products showed excellent properties. (2)Obtaining the data is still on the way. However some obtained data showed excellent value. (3)It is planned to carry out the neutron irradiation tests using High Flux Isotope Reactor at Oak Ridge National Laboratory up to the neutron fluence of 30 dpa, and the irradiation temperatures of 300-900℃. It is scheduled to end in 2014. The dimensional changes, elastic modulus, coefficient of thermal expansion, etc., will be studied.

INTRODUCTION

Very High Temperature Reactor (VHTR) of Generation-IV reactor is an advanced High Temperature Gas-cooled Reactor (HTGR) which is graphite-moderated and helium gas-cooled. It can provide high temperature helium gas about 950 °C to the reactor outlet. It is possible to use this high temperature as heat source not only for power generation but also for hydrogen production.

Graphite component used as core structure has high heat-resistant and large thermal capacity. It is expected to prevent from a rapid temperature change at the accident. From the view point of economy, graphite component will need long service life. In that case, it is expected to decrease the reactor core exchange frequency and graphite waste.

Service life of the graphite component is determined by the safety margin between residual stress after neutron irradiation in the reactor core and stress limit (permissive stress) [1]. The residual stress caused by mainly creep deformation after irradiation. Fig.1 shows the relation between the residual stress and stress limits. Japanese draft standard for HTGR[2] determined that stress limits is calculated by minimum ultimate tensile or compressive strength (S_u) and safety rate. The S_u value of graphite component is determined by the following equation.

$$S_u = \overline{X} - \left(\kappa_1 + \frac{\kappa_2}{\sqrt{n}} \right) \sigma$$

S_u : Minimum ultimate strength (MPa)

$\overline{X}$: Average of strength of data obtained by material test (MPa)

σ : Standard deviation of strength data obtained by material test (MPa)

n : Number of strength data points

κ_1 : 2.326 (equivalent to 99% survival probability)

κ_2 : 1.645 (equivalent to confidence level of 95% for one side when the standard deviation of a sample was supposed same as a statistical population's standard deviation considering the sampling error)

This equation indicates that high strength graphite with small variation has high Su value. It is expected to contribute toward added design margins and an extension of the service life of components.

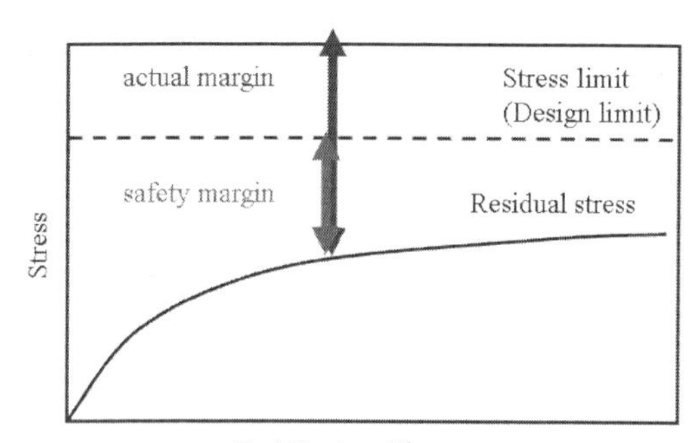

Fig.1 The relation of between the residual strength and stress limits

R&D PLAN and RESULT

Tokai Carbon Co. LTD. has started the development of nuclear grade graphite having high strength for the graphite component of VHTR. As the start of evaluation to the application for VHTR, the evaluation based on Japanese draft standard for HTGR[2] is advanced. Japanese draft standard demands as follows, (1)Guarantee a stable supply for long period, (2)Obtain the data for design, (3)Confirm stable features under high temperature and irradiation condition.

(1)Guarantee a stable supply for long period

To satisfy the requirement of stable supply, G347S and G458S grades were selected. G347S and G458S manufactured by Tokai Carbon Co. LTD. are fine-grained isotropic graphite. Fig.2 shows the texture of these materials. G458S is advanced product which has much crystallized area than G347S.

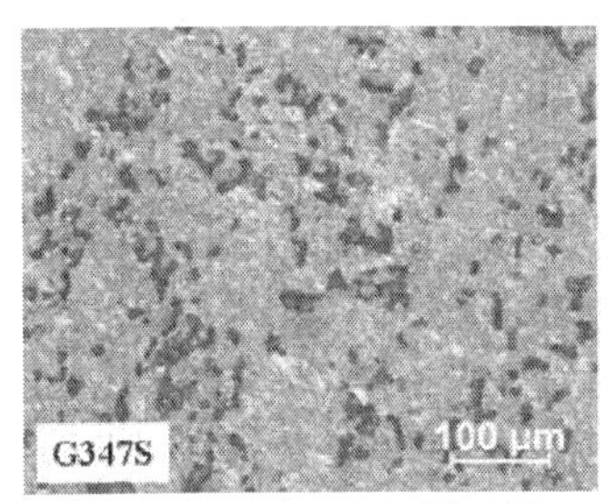

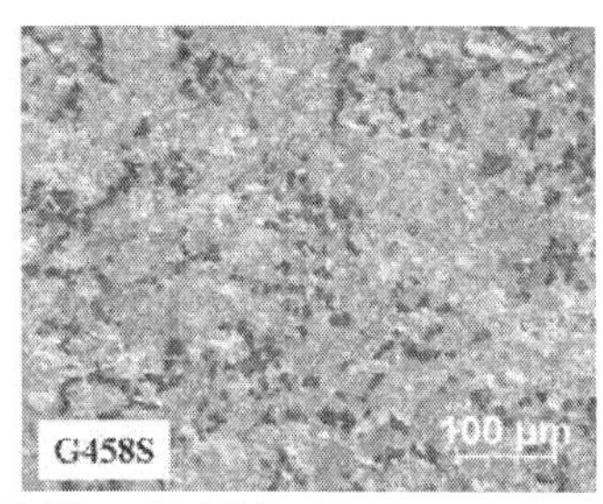

Fig.2 Texture of G347S and G458S

Table.1 shows the typical mechanical/thermal properties of G347S and G458S. Both materials have high strength and G458S has better thermal properties than G347S because of its high crystalline.

Table1 Mechanical/thermal properties of G347S and G458S

Grade	Bulk density (g/cm^3)	Electrical resistivity ($\mu\Omega\cdot m$)	Flexural strength (MPa)	Coefficient of thermal expansion ($x10^{-6}/°C$)	Thermal conductivity ($W/m\cdot k$)	Young's modulus (GPa)
G347S	1.85	11	49	4.2^{*1} (5.5^{*2})	116	10.8
G458S	1.86	9.5	53.9	3.1^{*1}(4.4^{*2})	139	11.3

(2)Obtain the data for design

Tokai Carbon Co. LTD. advance to obtain the data for design in collaboration with JAEA. Typical properties, impurities and mechanical strength should satisfy the requirement as the design of nuclear reactor. Table.2 shows the typical properties of G347S and G458S conforming to Japanese Industrial

Standards (JIS) with the specific target of core components used in High Temperature engineering Test Reactor (HTTR[3]). It also shows the specification standard conforming to ASTM D7219. G347S and G458S show similar properties with conventional materials[4]. Obtaining the data conforming to ASTM is also advanced.

Table2 Typical properties of G347S and G458S and specific target

Properties	Specific target	ASTM D7219	G347S	G458S
Bulk density (g/cm^3)	≥ 1.74	≥ 1.7	1.85	1.86
Electrical resistivity (μΩ•m)	≤ 13.0	-	11	9.5
Coefficient of thermal expansion ($x10^{-6}$/°C)	3.4 ~ 4.6[*3]	3.4 ~ 5.5[*4]	4.2[*1] (5.5[*2])	3.1[*1] (4.4[*2])
Ash content (wt ppm)	≤ 100	≤ 300	< 5	< 5
Flexural strength (MPa)	≥ 31.4	≥ 35	49	53.9
Anisotropy factor	≤ 1.15	-	1.06[*5]	1.10[*5]

Table.3 shows the data of impurities. All inspection items of G347S and G458S have excellent value.

Table3 Impurities data of G347S and G458S

inspection items	unit	Specific target	ASTM D7219	G347S	G458S
B Equ[*6]	ppm	< 1	< 2	< 0.29	< 0.29
Si		If Ash contents is 90ppm or less, it's passed	recommended to measure	< 0.1	< 0.1
Fe				< 0.04	< 0.04
Al				< 0.08	< 0.08
Ni				< 0.1	< 0.1
V				< 0.07	< 0.07
Ca				< 0.04	< 0.04
Li		< 0.01		< 0.01	< 0.01
Ash		< 100		< 5	< 5

In place of the S_u value of tensile and compressive strength, the Weibull plot of flexural strength of G347S is shown in Fig.3. Flexural data of G347S shows very small variation. G458S has similar value. From the correlation of flexural strength to tensile or compressive strength, Weibull plot of tensile and compressive strength are expected to show small variation. G347S and G458S also have tensile strength greater than 30MPa. As for this, G347S and G458S are expected to show the high S_u value. Since tensile and compressive data are not completed yet, it is planning to carry out the tests and evaluate the Su of G347S and G458S.

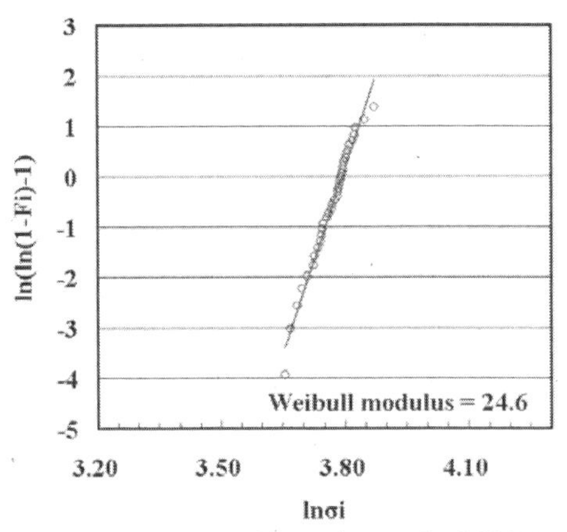

Fig.3 Weibull plot of flexural strength of G347S

(3) Confirms stable features under high temperature and irradiation condition

Tokai Carbon Co. LTD. commissioned Oak Ridge National Laboratory (ORNL) to perform neutron irradiation test using High Flux Isotope Reactor (HFIR). Fig.4 shows the plan of irradiation condition.

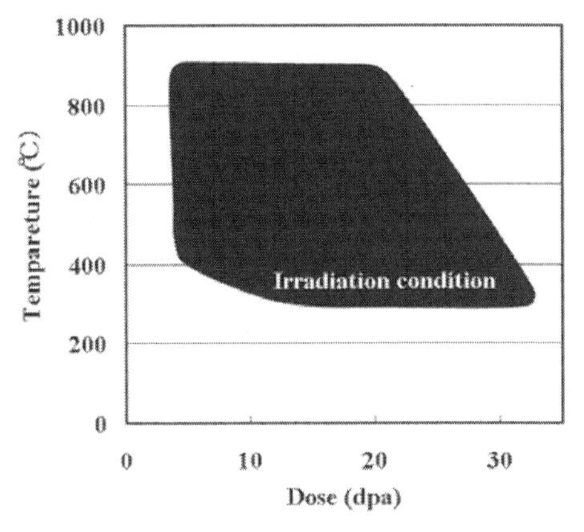

Fig.4 Irradiation test condition

It is planning to carry out the irradiation test up to the neutron fluence of 30 dpa, and the irradiation temperatures of 300-900°C. It is scheduled to end in 2014. The dimensional changes, elastic modulus, coefficient of thermal expansion, etc., will be studied.

SCHEDULE

Fig.5 shows a schedule of about 2 years in the future. As for irradiated characterization, it is scheduled to obtain the mechanical/thermal properties by early 2012. Evaluation the applicability for VHTR is advanced in parallel with them. These results are feedback to material or manufacturing improvement. As for irradiated characterization, the first irradiated data is obtained in a second half of 2012.

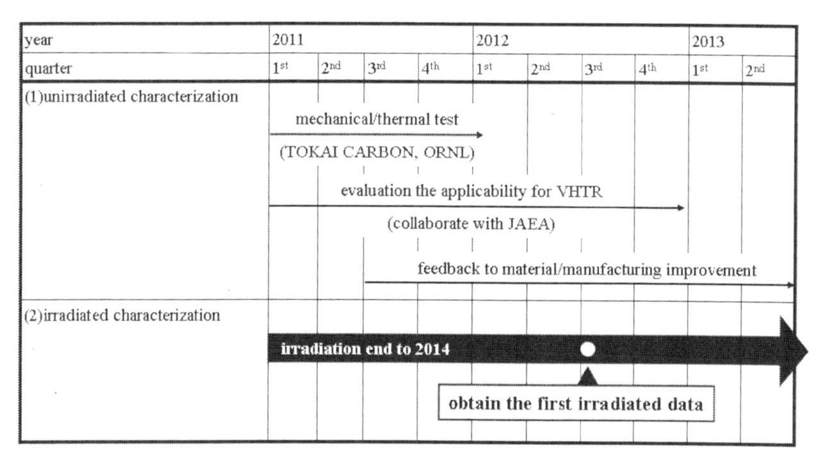

Fig.5 Schedule

CONCLUDING REMARKS

Tokai Carbon Co. LTD. has developed fine-grained isotropic nuclear graphite "G347S" and "G458S" having high strength to satisfy Japanese draft standard. G347 and G458S are expected to show an excellent data for VHTR design. Completing the data set for design is advanced. Irradiation test using HFIR at ORNL is scheduled to end in 2014 (~30dpa, 300-900°C).

FOOTNOTES

*1 RT~100°C

*2 RT~1000°C

*3 RT~400°C
*4 25~500°C
*5 Ratio of resistivity
*6 B Equivalent = B+1.57·Gd+0.941·Sm+0.855·Cd (HTTR)

REFERENCES

[1] T.Oku, M.Ishihara, Lifetime evaluation of graphite components for HTGRs, Nuclear Engineering and Design (2004), 227, 2, p209-217.
[2] T.Shibata, M.Eto, E.Kunimoto, S.Shiozawa, K.Sawa, T.Oku, T.Maruyama, Draft of Standard for Graphite Core Components in High Temperature Gas-cooled Reactor, JAEA-Research 2009-042(2010)
[3] S.Saito, et al., Design of High Temperature Engineering Test Reactor (HTTR), JAERI 1332 (1994).
[4] J.Sumita, T.Shibata, S.Hanawa, M.Ishihara, T.Iyoku, K.Sawa, Characteristics of First Loaded IG-110 Graphite in HTTR core, JAEA-Technology 2006-048(2006)

Crystalline, Amorphous, and Composite Materials for Waste Immobilization

FUNCTIONALIZED SILICA AEROGELS: ADVANCED MATERIALS TO CAPTURE AND IMMOBILIZE RADIOACTIVE IODINE

J. Matyáš, G.E. Fryxell, B.J. Busche, K. Wallace, and L.S. Fifield
Pacific Northwest National Laboratory
Richland, WA, USA

ABSTRACT

To support the future expansion of nuclear energy, an effective method is needed to capture and safely store radiological iodine-129 released during reprocessing of spent nuclear fuel. Various materials have been investigated to capture and immobilize iodine. In most cases, however, the materials that are effective for capturing iodine cannot subsequently be sintered/densified to create a stable composite that could be a viable waste form. We have developed chemically modified, highly porous, silica aerogels that show sorption capacities greater than 440 mg of I_2 per gram at 150 °C. An iodine uptake test in dry air with a concentration of iodine 44 mg/m^3 demonstrated no breakthrough after 3.5 h and indicated a decontamination factor in excess of 310. Preliminary densification tests showed that the I_2-loaded aerogels retained more than 92 mass% of I_2 after thermal sintering with pressure assistance at 1200 °C for 30 min. These high capture and retention efficiencies for I_2 can be further improved by optimizing the functionalization process and the chemistry as well as the sintering conditions.

INTRODUCTION

The difficulties encountered with opening Yucca Mountain (the site north of Las Vegas, Nevada, that was proposed for a nuclear waste repository) have led the U.S. Department of Energy (DOE) to seek strategies that would significantly delay, or even eliminate, the need for further geological disposal sites.[1] All of the realistic approaches suggest that the reprocessing of spent nuclear fuel could be part of a new strategy. As part of fuel reprocessing, chemically similar waste components, e.g., uranium, actinides, and long-lived fission products, need to be recovered, thereby enabling each waste stream to be immobilized separately or in combination with the others. The specifically tailored waste forms can then considerably reduce the volume of high-level nuclear waste.[2]

A significant challenge to meet the regulatory control requirements for radioactive discharges to the environment is the release of iodine-129 (^{129}I) during reprocessing. Other radioisotopes of iodine (^{131}I to ^{135}I) that are also produced in nuclear fuel have relatively short half-lifes, e.g, half-life of ^{131}I is 8 days, and decay to negligible levels by the time the fuel is ready to be processed. Because of its long half-life, 1.57×10^7 y, any ^{129}I emitted from a reprocessing plant will persist in the environment for tens of millions of years, and, although its activity is low, the dose associated with ^{129}I would gradually increase the longer these releases are allowed. In addition, the iodine, as a gas, is highly mobile in the environment. For these reasons, iodine must be efficiently captured and immobilized to meet the goal of limited emissions and the disposal-restriction regulations for radioactive waste.[3]

Iodine is primarily released during dissolution of spent nuclear fuel in nitric acid, with most of the iodine being carried in the dissolver off-gas. However, a portion is carried downstream in the dissolver solution, entering the solvent extraction operations or being released as a solid with the dissolver residues. One process under consideration is voloxidation (heating of spent nuclear fuel in atmosphere containing air, oxygen, ozone, or combination of these gases) prior to dissolution.[4] These two processes will result in essentially complete release of iodine in the dissolver and voloxidation off-gas streams.

Various wet scrubbing techniques or solid adsorbent methods have been investigated for removal of iodine from off-gas streams. Solid iodine adsorbents are preferred over liquid scrubbing because they remove iodine more efficiently and do not require the highly corrosive liquids typically associated with liquid scrubbers. In addition, adsorption columns filled with solid sorbent are generally simpler in design, resulting in a more reliable system with lower maintenance costs and fewer unit operations. The primary adsorbent materials available for radioiodine capture from off-gas include activated charcoal, macroreticular resins, and silver-based adsorbents. Activated charcoal is not considered an option for radioiodine removal from dissolver off-gas because of poor iodine retention at high temperatures (~150 °C) and a relatively low ignition point.[5] Also, large quantities of NO_2 in the off-gas may form unstable or explosive compounds in the carbon bed.[6] Macroreticular resins (functionalized acrylic esters or polystyrene with di-vinyl benzene cross-linking) are reasonably stable in feed streams containing NO_x and have an I_2capacity of 200 to 1000 mg /g, but their efficiency is greatly reduced at temperatures above 50 °C.[7] The silver-loaded adsorbents, such as silver-exchanged or silver-impregnated zeolites (mordenite and faujasite) as well as silica or alumina gels have been the most widely studied and are preferred for iodine removal because of their 1) effective retention of elemental and organic iodine, 2) high I_2 adsorption capacity of 80 to 235 mg /g, 3) high removal efficiencies, 4) high resistance to NO_x (except for silver-exchanged faujasite, [AgX]), and 5) non-flammability.

The silver-exchanged (AgZ) or silver-reduced (Ag^0Z) mordenites are considered the benchmark for radiological iodine capture, and it is believed that their performance can be further improved by understanding the distribution and properties of silver iodide and the role of the zeolite support.[8] These materials, however, cannot be subsequently sintered or densified to create a composite that could be a viable waste form. Potential waste forms for iodine are based either on 1) encapsulating iodine-loaded silver-mordenite in low-melting glass[9], cement[10], or silico-geopolymer[11], or 2) removing iodine from the silver-mordenite and subsequently encapsulating it in apatite-like minerals, low-melting glass, bismuth-containing ceramics, and iodide or iodate sodalites.[12-14] The process of encapsulating iodine-loaded silver-mordenite in low-melting, bismuth-zinc-borate glass is currently under development.[15]

At Pacific Northwest National Laboratory (PNNL), we are developing advanced, high-surface-area, functionalized silica aerogels with high I_2-loading and retention capacities that show great promise for removing and immobilizing ^{129}I from an advanced fuel cycle reprocessing plant. In this paper, we present and discuss the results of proof-of-concept characterization studies of our developed material.

MATERIALS AND METHODS

Materials

To avoid the time-consuming, sol-gel synthesis, our efforts focused on testing and functionalizing commercially available silica aerogel materials from United Nuclear (Laingburgh, MI). The 3-mercaptopropyltrimethoxysilane (3-MPTMS, 95%), silver nitrate ($AgNO_3$, 99%), and isopropanol (IPA, 99%) were purchased from Sigma-Aldrich (St. Louis, MO) and used for thiol functionalities and silver impregnation, respectively. A cylinder of 2.7% H_2 balanced with Ar was purchased from Oxarc (Pasco, WA) and used in the silver reduction step. Elemental I_2 for the I_2 uptake test was generated with a permeation tube from VICI Metronics (Poulsbo, WA). The permeation tube had the National Institute of Standards and Technology (N.I.S.T) certified I_2 permeation rate of 1371.58 ng/min at 100 °C.

Analytical Methods

Brunauer, Emmett, Teller (BET) gas adsorption analysis, simultaneous thermogravimety and differential scanning calorimetry (TGA-DSC), and transmission electron microscopy (TEM) were used to characterize samples from different stages in the development of functionalized silica aerogels. Inductively coupled plasma-mass spectrometry (ICP-MS) was used to determine the concentrations of iodine in the caustic solutions and sorbent.

The BET analysis was performed on samples degassed at 25-80 °C under vacuum. The data from nitrogen adsorption/desorption at 77 K were collected with a Quantachrome Autosorb 6-B gas sorption system (Quantachrome Instruments, Boynton Beach, FL). The surface area was determined from the isotherm with the five-point BET method. The Barrett–Joyner–Halenda (BJH) method was used to calculate the pore volume and average pore size. The TGA-DSC analyses were performed with a TA Instruments SDT Q-600 (TA Instruments, New Castle, DE) on small samples (6-25 mg) heated at 5 °C/min from room temperature to 1000 °C under flowing air, 20 mL/min at 25 °C. For TEM, an aerogel granule was embedded in epoxy and cured at 60 °C for 24 h; a small section was cleaved with a microtome knife and then deposited on the copper TEM grids sputtered with carbon. The sample was analyzed with a Tecnai T-12 TEM (FEI, Hillsboro, OR) operating at 120 kV equipped with LaB_6 filament. Images were collected with 2×2K Ultrascan 1000 charge-coupled device (Gatan Inc., Pleasanton, CA) calibrated to the TEM camera length to enable direct measurements correlated with the magnification of the acquired images. For the iodine analyses with ICP-MS, caustic solutions were analyzed directly; solid samples were first fused with sodium carbonate to obtain solutions for analyses.

Synthesis of functionalized silica aerogels

The surfaces of free-standing aerogels were chemically modified with the supercritical fluid chemistry previously developed at PNNL.[16] This procedure is both fast and efficient, produces very little waste, making it a very "green" process, and is suitable as a candidate for ultimately manufacturing these materials (similar PNNL technology has been licensed to Steward Environmental Solutions, LLC, and is currently being used for commercial sorbent material production).

The formation of silver nanoparticles on silica-aerogel surfaces is shown schematically in Figure 1. The hydroetched and hydrated granules of silica aerogel were functionalized by the deposition of a propylthiol monolayer on the surface by treating the aerogel with 3-MPTMS in supercritical CO_2 solvent (52 MPa and 150 °C). A layer of Ag(I) ions was installed on the thiol monolayer interface by treatment with 200 mL of 3.25 mass% $AgNO_3$ solution. About 25 mL of isopropanol (IPA) was added to the solution to facilitate wetting of the moderately hydrophobic thiol-modified aerogel, which led to the complete dispersion of coated aerogel granules. The silver nanoparticles were produced on the silica aerogel pore surfaces by reducing the silver thiolate adduct ions at 165 °C for 2 h under flowing 2.7% H_2 in Ar.

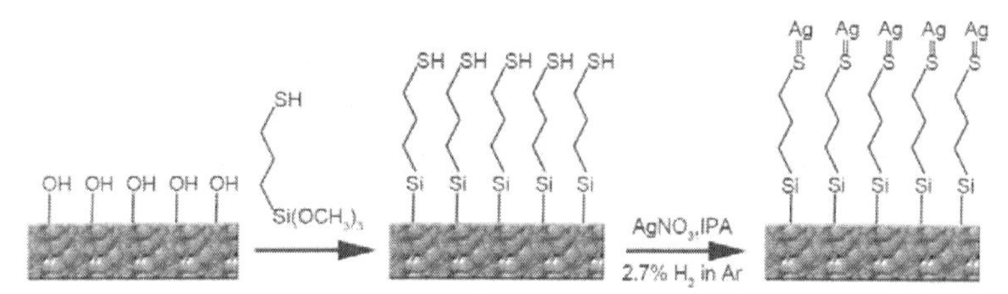

Figure 1. The reaction path to form silver nanoparticles on silica aerogel surfaces.

Maximum sorption test

Figure 2 shows a schematic of the experimental apparatus that was designed and assembled to determine the maximum sorption capacity of functionalized silica aerogel for I_2. A Teflon® vessel with ~1 g of solid I_2 was connected to the column with sorbent, and the whole assembly was heated in an oven to 150 °C. The I_2 gas from subliming I_2 was allowed to pass through the column with a sorbent to the beaker containing 1 M NaOH over a period of 24 h. To remove the loosely bonded or excess I_2, the column was detached, heat-treated in another oven at 150 °C overnight, and transferred into a desiccator under vacuum (- 91.4 kPa) for 24 h. Subsequently, the column was removed from the desiccator and weighed with an analytical balance with 0.1 mg sensitivity to obtain the mass gain (I_2 sorption capacity in mg/g) of Ag-reduced aerogel.

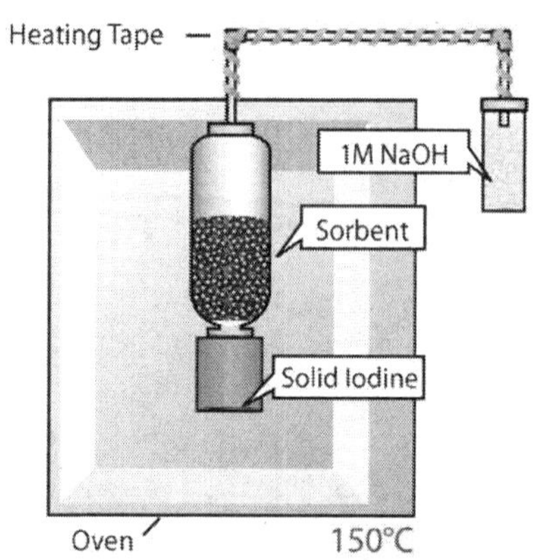

Figure 2. A schematic of the experimental apparatus for determining the maximum sorption capacity of functionalized silica aerogels.

Iodine uptake test in air containing a low concentration of iodine

Figure 3 shows a schematic diagram of the experimental apparatus used to evaluate I_2 uptake at low concentrations. Breathing air with a H_2O concentration of 50 mg/m^3 was allowed to pass through an U-tube containing glass beads and an I_2 permeation tube. The iodine concentration of 44 mg/m^3 in the carrier gas was established by heating the permeation tube in the oven to 100 °C and adjusting the carrier gas flow rate to 18.9 mL/min (linear velocity 401 mm/s). Two independent 1-h long calibration runs (no sorbent present in the column) were performed to confirm the concentration of I_2 in the carrier gas. Concentrations of I_2 in the solutions were 2468 and 2495 μg/L and agreed well with 44 mg/m^3 of iodine in the carrier gas. The air with iodine was allowed to pass through the column with 537 mg of Ag-reduced aerogel (heated in the oven to 150 °C) to the beaker containing 20 mL of 0.1 M NaOH. The beaker with caustic solution was replaced every 30 min. The collected solutions were analyzed for I_2 with inductively coupled plasma-mass spectrometry (ICP-MS) (detection limit for I_2 was 4 μg/L).

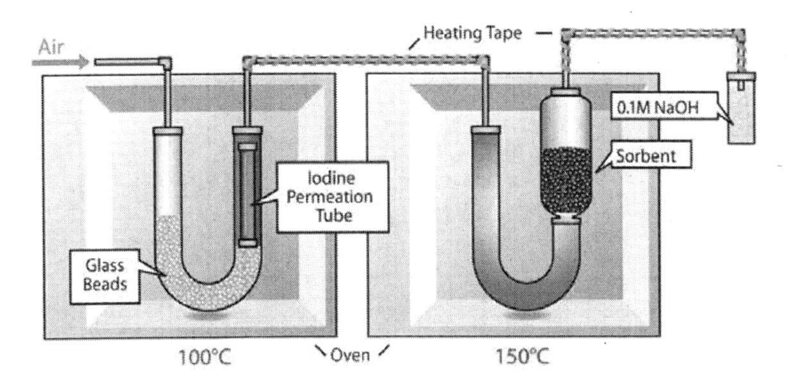

Figure 3. Schematic diagram of the experimental set-up for iodine uptake test in breathing air containing 4.2 ppm of iodine.

Immobilization of iodine-loaded silica aerogel

Thermal sintering with pressure assistance was used to effectively encapsulate the iodine gas in a durable silica glass matrix. Iodine-loaded aerogel (502 mg) with an I_2 concentration of 154.2 mg/g (as determined with ICP-MS) was covered with a thin quartz lid and pressed with 103.4 kPa pressure inside the alumina crucible at 1200 °C for 30 min. Then, the sample was removed from the crucible and analyzed for iodine with ICP-MS.

RESULTS AND DISCUSSION

Characterization

The high degree of surface functionalization was demonstrated and characterized with BET gas adsorption analysis, TGA-DSC, and TEM. The BET surface areas decreased from 818 m^2/g for raw aerogel to 620 m^2/g for thiol-coated aerogel and to 199.1 m^2/g for Ag-functionalized aerogel. This decrease was consistent with the addition of mass and the occupation of pore volume. The pore volume followed the same trend and decreased from 5.4 to 3.5 and 0.18 cm^3/g, respectively, confirming a good surface coating with functionalized groups. Figure 4 shows a TEM image of Ag-functionalized aerogel. TEM analyses revealed a uniform distribution and size of the silver nanoparticles (< 10 nm) on the surfaces of a highly porous silica aerogel sample.

Figure 5 shows TGA curves for raw and thiol-aerogels. The raw aerogel exhibited two distinct mass losses during the heating. First, 1.8 mass% of physically adsorbed water was lost at temperatures below 100 °C, and then 9.4 mass% loss of silanol groups gradually disappeared at temperatures above 350 °C. For thiol-aerogel, an initial 0.4 mass% loss between 25 °C and 50 °C was attributed to the residual IPA solution physically adsorbed to the monolayer. The mass loss of 41.3 mass% between 50 °C and 530 °C was from the decomposition of the propylthiol monolayer anchored to the aerogel pore surfaces. The mass loss of 2.5 mass% at temperatures above 530 °C corresponded to the removal of the remaining silanol groups. Figure 6 shows TGA-DSC curves for two aerogels, Ag-impregnated (aerogel after treatment with $AgNO_3$) and Ag-reduced (aerogel after reduction under H_2/Ar). Both aerogels showed mass loss of ~ 8 mass% in the temperature range of 265 °C to 400 °C from the decomposition of propylthiol monolayer.

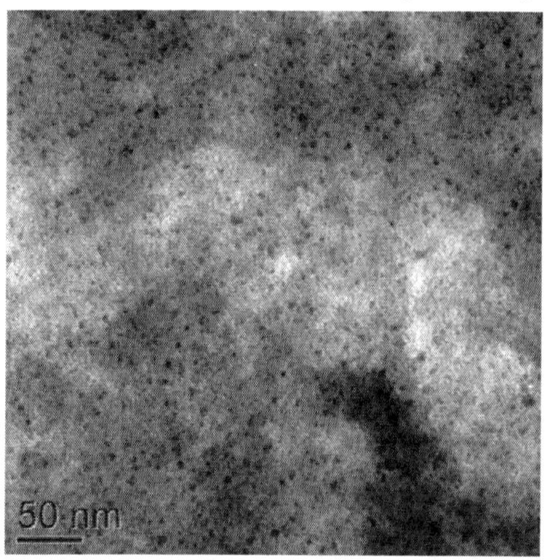

Figure 4. A TEM image of Ag-functionalized aerogel (silver nanoparticles are black dots).

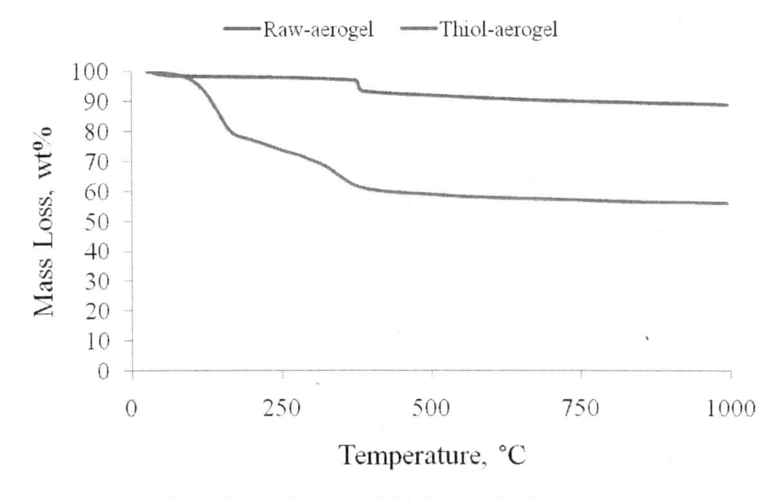

Figure 5. Results of mass losses for raw and thiol-aerogels with increasing temperature as determined with TGA.

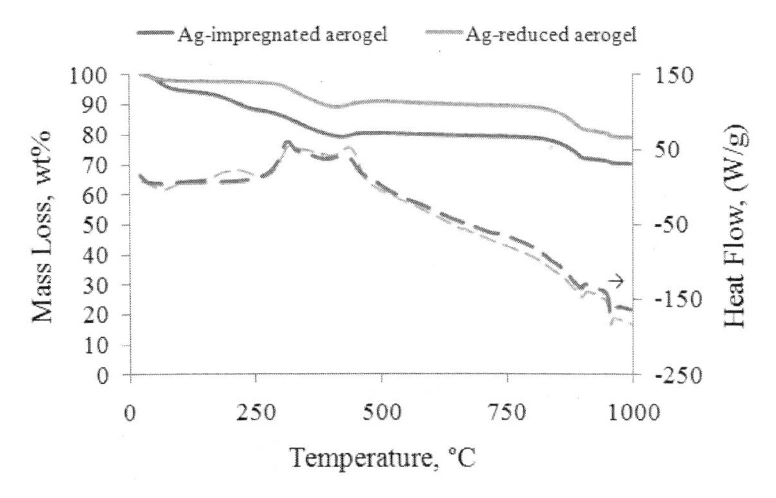

Figure 6. Results from TGA-DSC measurements for Ag-impregnated and reduced aerogels.

The sample from the maximum sorption test gained 55.9 mass%, suggesting a high sorption capacity of I_2-loaded functionalized aerogel. This was consistent with the TGA-DSC results (Figure 7). Also shown is the mass-corrected heat flow as a function of temperature. The total sample mass lost was 56.8 mass%. The majority, 44.3 mass%, comes from the decomposition of silver iodide that starts at 558 °C,[17] corresponding to the maximum sorption capacity of aerogel for iodine. The remaining 12.5 mass% includes mass losses from physically absorbed water, a propylthiol monolayer, and physically sorbed iodine.

The analysis of NaOH solutions that were collected during the I_2 uptake test did not detect any I_2 (I_2 concentrations were below the detection limit of ICP-MS), indicating 100% capture efficiency and corresponding to a decontamination factor of >310. Functionalized silica aerogel removed 175 μg of I_2 out of 175 μg possible.

Figure 8 shows a photograph of the I_2-loaded silica aerogel after thermal sintering with pressure assistance. The concentration of I_2 in the glass sample, as determined with ICP-MS, was 142.2 mg/g. Since the concentration of I_2 in the I_2-loaded aerogel before sintering was 154.2 mg/g, we were able to retain 92.2% of the sorbed I_2 in the final product.

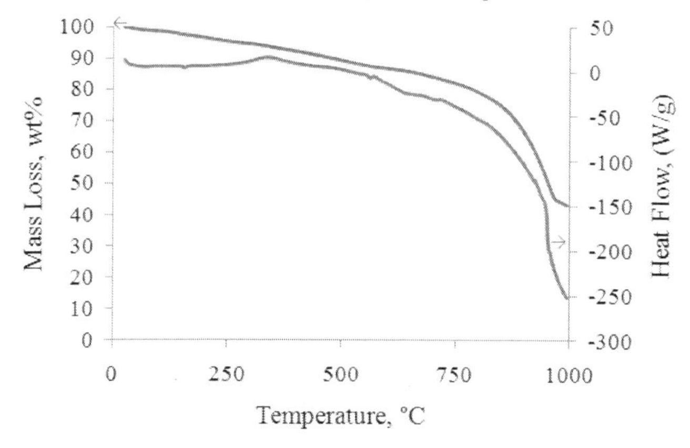

Figure 7. Results from the TGA-DSC analyses of iodine-loaded aerogel.

Figure 8. A photograph of the I_2-loaded silica aerogel in an alumina crucible (the yellow tint is an artifact of the photographic process; the crucible remained white) after thermal sintering at 1200 °C for 30 min 103.4 kPa pressure.

CONCLUSION

This work has shown that it is possible to functionalize aerogels, something that is difficult to do with traditional solvent-based methods because of the frailty of the aerogel backbone. These functionalized aerogels can be effective for capturing a variety of airborne target species and, once laden with contaminants, these materials can be sintered or densified. At the laboratory scale, we have demonstrated that Ag-functionalized silica aerogels are capable of efficient capture and sequestration of gaseous I_2. The functionalized aerogels exhibit a high sorption capacity for I_2 of more than 440 mg/g at 150 °C with a decontamination factor of >310 and retained > 90% of sorbed I_2 after thermal sintering with pressure assistance at 1200 °C for 30 min.

In the next phase of the project, we would like to further improve the capture efficiencies of functionalized aerogels for I_2 as well as the retention of I_2 in densified aerogels. In addition, we will investigate the waste form performance of densified aerogels with the product consistency test.

ACKNOWLEDGEMENTS

Authors are grateful to Xiaohong Shari Li for BET analysis and Alice Dohnalkova for TEM investigations. This work was funded by the U.S. Department of Energy's Fuel Cycle Research and Development Program. PNNL is operated for DOE by Battelle under Contract DE-AC05-76RL01830.

REFERENCES

[1]E. A. Schneider, M. R. Deinert, and K. B. Cady, Cost analysis of the US spent nuclear fuel reprocessing facility, *Energy Economics*, **31** (5), 627-634 (2009).

[2]B. D. Begg, R. A. Day, S. A. Moricca, M. W. A. Stewart, E. R. Vance, R. Muir, Low-risk waste forms to lock up high-level nuclear waste, WM'05, Tucson, AZ (2005).

[3]40 CFR 190, Environmental Radiation Protection Standards for Nuclear Power Operations, Code of Federal Regulations, *Environmental Protection Agency*.

[4]D. Haefner, J. Law, T. Tranter, System design and description and requirements for modeling the off-gas systems for fuel recycling facilities, INL/EXT-10-18845, Idaho National Laboratory (2010).

[5]*International Atomic Energy Agency*, Design of off-gas and air cleaning systems at nuclear power plants, IAEA-TECDOC-274, IAEA, Vienna (1987).

[6]D.T. Pence, F.A. Duce, and W.J. Maeck, Iodine Adsorbents Program, Idaho Chemical Programs Annual Technical Report Fiscal Year 1971, ICP-1006, Idaho National Laboratory (1972).

[7]D.R. Haefner and T.J. Tranter, Methods of Gas Phase Capture of Iodine from Fuel Reprocessing Off-Gas: A Literature survey, INL/EXT-07-12299, Idaho National Laboratory (2007).

[8]K. W. Chapman, P. J. Chupas, and T. M. Nenoff, Radioactive iodine capture in silver-containing mordenites through nanoscale silver iodide formation, *J. Am. Chem. Soc.*, **132** (26), 8897-8899 (2010).

[9]T. J. Garino, T. M. Nenoff, J. L. Krumhansl, and D. X. Rademacher, Development of waste forms for radioactive iodine, in Ceramics for Environmental and Energy Applications (eds A. Boccaccini, J. Marra, F. Dogan, H.-T. Lin, T. Watanabe and M. Singh), John Wiley & Sons, Inc., Hoboken, NJ, USA. doi: 10.1002/9780470909874.ch5 (2010).

[10]R. D. Scheele, C. F. Wend, W. C. Buchmiller, A. E. Kozelisky, R. L. Sell, Preliminary evaluation of spent silver mordenite disposal forms resulting from gaseous radioiodine control at Hanford's Waste Treatment Plant, PNWD-3225, WTP-RPT-039 (2002).
[11]W. Gong, W. Lutze, I. L. Pegg. Low-temperature solidification of radioactive and hazardous wastes, US Patent #7855313 (2010).
[12]M.W Barnes, D. M. Roy and L. D. Wakeley, Leaching of composites of cement with the radioactive waste forms strontium powellite and iodine sodalite at 90 °C, in Advances in Ceramics, eds. GG Wicks and WA Ross, **8**, 406-12. American Ceramic Society, Columbus, OH (1983).
[13]E. R. Vance, White W.B. and Roy R., Synthetic Mineral Phases for the Fixation of I-129. *American Ceramic Society Bulletin*, **59**, 397-97 (1980).
[14]D. M Strachan. and H. Babad, Iodide and Iodate Sodalites for Long-Term Storage of I-129. *American Ceramic Society Bulletin*, **58**, 327-27 (1979).
[15]T. M. Nenoff, J. L. Krumhansl, T. Garino, N. W. Ockwig and K. Holt-Larese, Waste Form Development and Testing for I_2. GNEP-WAST-PMO-MI-DV-2008-000149, U.S. Department of Energy, Office of Nuclear Energy, Washington, D.C. (2008).
[16]T. S. Zemanian, G. E. Fryxell, and O.A Ustyugov, Monolayer coated aerogels and method of making, US Patent #7019037 (2006).
[17]M. Reháková, A. Sopková, K. Jesenák, and V. Š. Fajnor, Thermal analysis of the synthetic zeolite ZSM5 and its silver iodide form, *Journal of Thermal Analysis and Calorimetry*, 50 (3), 505-509 (1997).

LAYERED DOUBLE HYDROXIDES FOR ANION CAPTURE AND STORAGE

J. Phillips[1]; L. J. Vandeperre[1]
[1]Department of Materials, Centre for Advanced Structural Ceramics
Imperial College London, South Kensington Campus, London, SW7 2AZ

ABSTRACT

The technetium isotope, Tc, is a long lived radionuclide with a half life of $2.1x10^5$ years, produced with a yield of ~6% during nuclear fission which exists in solution as the pertechnetate anion, TcO_4^-. The pertechnetate anion does not bond well to soils and is highly mobile in groundwater as such it is highly desirable to capture and immobilise the TcO_4^- anion for long term storage. An integrated solution is proposed for the capture and disposal of this anion using layered double hydroxides (LDHs) as the capture medium. To this end, LDHs with the general formula $Ca_{(1-x)}$ $(Al_{(1-y)},Fe_{(y)})_x$ $(OH)_2$ xNO_3 . nH_2O were produced via a co-precipitation method from nitrate salts. This composition was chosen as it is suitable for thermal conversion to Brownmillerite, $Ca_2(Al,Fe)_2O_5$, a phase found in cements. LDHs with composition $Ca_{0.67}$ $(Al_{(0.5)},Fe^{3+}{}_{(0.5)})_{0.33}$ $(OH)_2$ $0.33NO_3.nH_2O$ were produced and exposed to solutions containing either single or pairs of anions (NO_3^-,Cl^-,CO_3^{2-} and ReO_4^- as a surrogate for TcO_4^-). It was found that exchange occurred in each case and the preference for intercalation was $Cl \approx CO_3 > NO_3 > ReO_4$. LDHs containing Ca, Al, and Fe are therefore suitable phases for the capture of anions, the composition of the LDH can be tailored to produce materials suitable for thermal conversion to ceramic phases.

INTRODUCTION

Technetium (^{99}Tc) is produced during nuclear fission reactions with a yield of approximately 6% [1]. It is a long lived radionuclide with a half life of $2.1x10^5$ years; containment of ^{99}Tc is difficult due to its anionic nature. In elemental form technetium is relatively immobile, however it readily oxidises to the pertechnetate anion, TcO_4^-. The pertechnetate anion does not bind well with soils and is therefore highly mobile in groundwater. In the UK a process was developed to capture the pertechnetate anion from solution with Tetraphenylphosphonium Bromide, TPPB [2] during the treatment of reprocessing effluent. This allows the TcO_4^- anion to be filtered out in the same step as other radionuclides. The filtered solids are then incorporated into a cementitious waste form. It has been shown that in strongly alkaline environments such as those found within the pores of cements TPPB can degrade by alkaline hydrolysis [2], resulting in the release of the pertechnetate anion.

An integrated capture and disposal method is proposed for pertechnetate and other anions using layered double hydroxides as the capture medium. The composition of the layered double hydroxide is selected such that after capture of the anion, the powder can be thermally converted to phases compatible with a cementitious environment or added directly to a cement composition.

Layered double hydroxides (LDH) are structurally similar to the mineral Brucite, $Mg(OH)_2$. Brucite consists of large two dimensional sheets of metal hydroxide, formed by edge sharing of M^{2+} cation octahedra. In the case of layered double hydroxides a fraction, x, of the divalent sites undergo an isomorphous substitution with a trivalent cation (Al^{3+}, Fe^{3+}) [3]. This generates a net positive charge on the surface of the metal hydroxide sheet; this charge is balanced by the incorporation of anions within the interlayer galleries. The potential compositional range for LDHs is broad, but is more limited for calcium based LDHs than for those based on magnesium for example due to the large ionic radius of calcium which causes cation ordering within the metal hydroxide sheet. The choice of interlayer anion is flexible and a broad selection of anions have been incorporated ranging from Cl^-, CO_3^{2-} or CrO_4^{2-} [4-7] to large organic molecules[8]. The general formula for the LDHs employed in this study is presented in Equation 1 below.

$$Ca_{(1-x)}\,(Al_{(1-y)},Fe_{(y)})_x\,(OH)_2\,A^{n-}_{x/n} \tag{1}$$

Other anions such as $^{36}Cl^-$ which may present a problem during the disposal of irradiated graphite wastes [9] may also be disposed of in layered double hydroxide materials. The small thermochemical radius of the chlorine anion [10] makes it an ideal candidate to be incorporated in to the interlayer of LDHs where high charge density is favourable. One calcium based LDH is known as Friedel's Salt [11]. It has the formula given in Equation 2. This chlorine intercalated calcium based LDH is stable which is beneficial for long term storage.

$$Ca_2Al(OH)_6(Cl,\,OH)\;2\;H_2O. \tag{2}$$

The long term stability of a pertechnetate intercalated LDH is untested and it is proposed that it may be possible to convert the LDH phase to a stable ceramic phase by heat treatment. It is expected from previous studies on calcium containing layered double hydroxides that the thermal decomposition will occur in a number of mass loss steps [11-13] and result in the formation of a mixed oxide phase. The compositions employed in this study are intended to convert to the Brownmillerite, Ca_2AlFeO_5, a phase commonly found in cements.

MATERIALS AND METHODS

A co-precipitation method was employed to produce the layered double hydroxide powders in this study. The nitrate salts of calcium, aluminium and iron ($Ca(NO_3)_2.4H_2O$, $Al(NO_3)_3.9H_2O$, $Fe(NO_3)_3.9H_2O$) were dissolved in distilled water in the desired stoichiometric ratio to a total concentration of 1 M. This solution was added dropwise to a solution maintained at $pH>12.5$ by the addition of 1 M NaOH and the sodium salt of the desired interlayer anion ($NaNO_3$, Na_2CO_3, or NaCl). The resulting suspension was allowed to age for 1 hour before the solids were removed by vacuum enhanced filtration, (0.45 μm Millipore) before re-suspension in distilled water to remove any undesired nitrate salts from the product. After multiple rinses, the filtered product was collected and dried at 80 °C in an air atmosphere, prior to being ground to a fine powder.

The powders were characterized by x-ray diffraction using a Philips PW1720 powder x-ray diffractometer and by combined Thermogravimetric and Differential Thermal Analysis with a Netzsch STA 449 F1. For the latter the powders were heated in an open alumina crucible in flowing air at a rate of 10 °C per minute.

To study the competitive uptake of anions from solution 1 g of nitrate intercalated LDH powder was added to a 50 ml centrifuge tube containing pairs of interlayer anions at 3 different molar ratios 0.1:0.9, 0.5:0.5 and 0.9:0.1 (corrected for charge differences of the anions). The relative preference for carbonate, chloride and nitrate anions was investigated as these could be competing anions for pertechnetate sorption. The potential for pertechnetate sorption was investigated using ReO_4^- as an inactive surrogate for TcO_4^- due to its similar thermochemical radius and chemical properties. Tests involving ReO_4 were not conducted in the competitive manner described above but by testing its capture from a 0.1M ReO_4^- single anion solution.

RESULTS AND DISCUSSION

Competitive adsorption of possible background anions

Figure 1a shows the (003) reflection obtained when a sample of $Ca_4AlFe(OH)_6(NO_3)_2.2H_2O$ was exposed to solutions containing pairs of anions ($Cl:CO_3$). This reflection is shown as it is a direct measure for the interlayer thickness. It is clear that the resulting LDH structure contains two distinct interlayer spacings and that there is a slight preference for the chloride anion over the carbonate anion. Figure 2b shows that prolonged exposure to a solution with a high concentration of CO_3^{2-} resulted in a loss of the LDH structure. Quantitative Rietveld analysis of the data allowed to establish that the order of preference for the uptake of anions was $Cl^- \approx CO_3^{2-} > NO_3^-$. The high selectivity for Cl^-, certainly when compared to Mg-based LDHs, suggests that these materials might be useful for capture of ^{36}Cl from graphitic wastes. The deleterious effect of prolonged exposure to carbonate rich solutions is evident from Figure 1b where the LDH structure has been lost, further details can be found in [14].

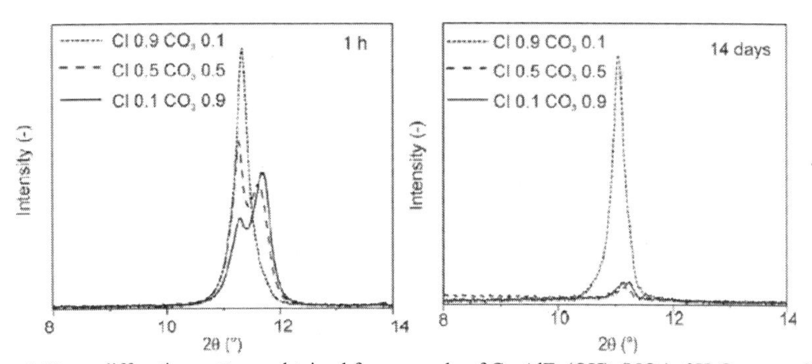

Figure 1 X-ray diffraction patterns obtained for a sample of $Ca_4AlFe(OH)_6(NO_3)_2.2H_2O$ exposed to a solution containing Cl and CO_3 anions in a range of molar ratios .

Reprinted from [14] with permission from Elsevier.

The thermal analyses of the calcium aluminium iron LDH which contained different anions each demonstrated that the weight losses occur in a number of steps as shown in Figures 2-4. The mass losses for each step are attributable to the loss of interlayer water below 200 °C, followed by the gradual dehydroxylation of the metal sheets and finally the final volatilization of the interlayer anion or a complex formed during decomposition. The small additional weight loss step at ~700 °C is tentatively attributed to the decomposition of $Ca(OH)_2$ forming CaO and a H_2O molecule [15, 16]. For all materials studied, the overall weight loss was consistent with 2 water molecules incorporated in the structural formula. Given that the materials were dried at 80 °C, this is consistent with the findings of Renaudin et al., which found that the number of water molecules can associated with each calcium atom was 1 at room temperature and 0.5 at 70 °C [12].

In the case of the chlorine intercalated LDH, the chlorine volatilises in the form of HCl at ~500 °C. This may occur through an intermediary phase such as CaOHCl, formed during the collapse of the LDH structure. The other mass loss events remain the same as for the nitrate intercalated LDH. The decomposition temperatures are in good agreement with the formation of calcium compounds formed between the metal hydroxide sheet and the interlayer anion, such as $CaCO_3$ or CaOHCl.

Adsorption of ReO_4^- as a surrogate for TcO_4-

Each of the samples tested with ReO_4^- resulted in the formation of an LDH with an interlayer spacing attributable to a carbonate intercalated LDH and the trace obtained from thermal analysis of the samples after exchange with ReO_4^- containing solutions was also consistent with a carbonate intercalated LDH. The only carbonate ions present are due to natural carbonation of the water used and hence the affinity for ReO_4^- appears to be limited and in any case much smaller than for CO_3^{2-}. Further tests with decarbonated water under inert gas cover are planned to establish the ability for ReO_4^- uptake further.

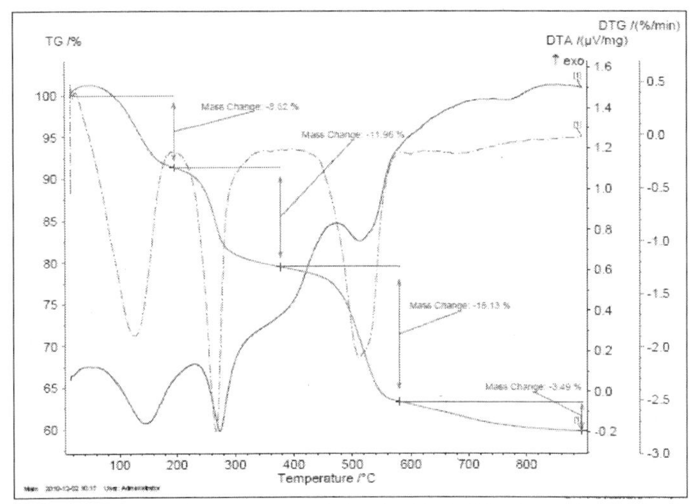

Figure 2 Thermogravimetric analysis (with first derivative (dashed line)) and differential thermal analysis of $Ca_2Al,Fe^{3+}(OH)_6(NO_3)_2.2\ H_2O$

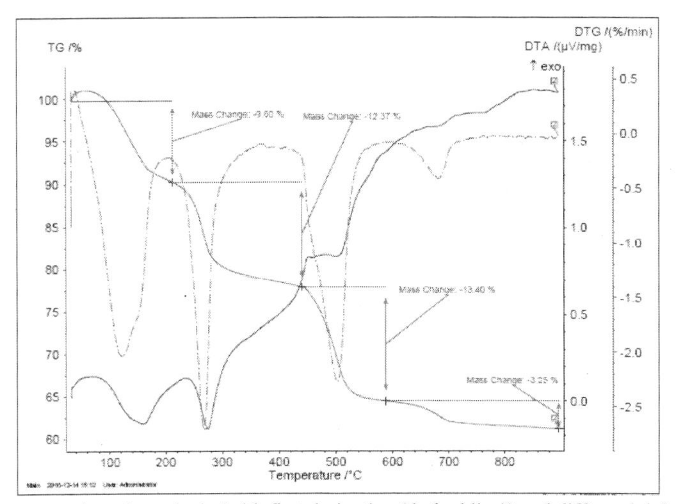

Figure 3 Thermogravimetric analysis (with first derivative (dashed line)) and differential thermal analysis of $Ca_2Al,Fe^{3+}(OH)_6(Cl)_2.2\ H_2O$

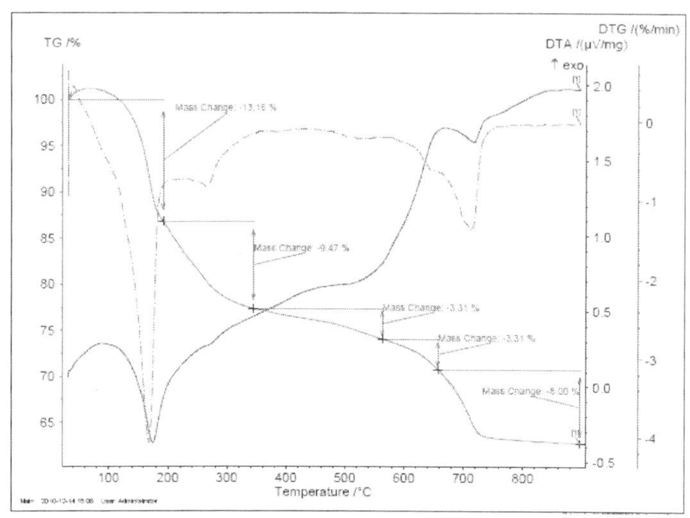

Figure 4 Thermogravimetric analysis (with first derivative (dashed line)) and differential thermal analysis of $Ca_2Al,Fe^{3+}(OH)_6CO_3.2\ H_2O$

CONCLUSIONS

Layered double hydroxide phases with a ternary composition of calcium aluminium and iron were produced by a co-precipitation method. The relative preference for potentially competing anions was established as $Cl^- \approx CO_3^{2-} > NO_3^-$. Initial trials to capture ReO_4^- were not successful and indicated that competition from the carbonate anion might make pertechnetate incorporation difficult if not impossible. The high affinity for both chloride and carbonate might be advantageous in cleaning graphitic wastes.

Thermal analysis showed that each of the LDH powders had two water molecules within the unit cell. The weight losses of the ternary LDHs were in good agreement with literature values for losses in simple binary systems. The coordination of anions in the interlayer with the calcium cation in the hydroxide sheet may have caused the preferential formation of phases such as $CaCO_3$ or CaOHCl during decomposition.

ACKNOWLEDGEMENTS

The authors would like to acknowledge funding from the United Kingdom's Engineering and Physical Sciences Research Council via the Decommissioning Immobilisation and Management of Nuclear Waste for Disposal (DIAMOND) Consortium.

REFERENCES

[1]J. Reed, Technetium to go, Nuclear Engineering International, 49 (2004) 14-17.
[2]S. Aldridge, P. Warwick, N. Evans, S. Vines, Degradation of tetraphenylphosphonium bromide at high pH and its effect on radionuclide solubility, Chemosphere, 66 (2007) 672-676.
[3]I. Rousselot, C. Taviot-Guého, F. Leroux, P. Léone, P. Palvadeau, J.-P. Besse, Insights on the Structural Chemistry of Hydrocalumite and Hydrotalcite-like Materials: Investigation of the Series $Ca_2M^{3+}(OH)_6Cl{\cdot}2H_2O$ (M^{3+}: Al^{3+}, Ga^{3+}, Fe^{3+}, and Sc^{3+}) by X-Ray Powder Diffraction, Journal of Solid State Chemistry, 167 (2002) 137-144.
[4]G. Renaudin, M. Francois, The lamellar double-hydroxide (LDH) compound with composition 3 Ca O . Al_2O_3 . $Ca(NO_3)_2$. 10 H_2O, Acta Crystallographica C, 55 (1999) 835-838.
[5]G. Renaudin, M. Francois, O. Evrard, Order and disorder in the lamellar hydrated tetracalcium monocarboaluminate compound, Cement and Concrete Research, 29 (1999) 63-69.
[6]G. Renaudin, F. Kubel, J.P. Rivera, M. Francois, Structural phase transition and high temperature phase structure of Friedels salt, 3(CaO) . (Al_2 O_3) . ($CaCl_2$) . 10(H_2O), Cement and Concrete Research, 29 (1999) 1937-1942.
[7]S.V. Prasanna, P.V. Kamath, C. Shivakumara, Synthesis and characterization of layered double hydroxides (LDHs) with intercalated chromate ions, Materials Research Bulletin, 42 (2007) 1028-1039.
[8]Y. Seida, Y. Nakano, Removal of humic substances by layered double hydroxide containing iron, Water Research, 34 (2000) 1487-1494.
[9]F.J. Brown, J.D. Palmer, P. Wood, Derivation of a radionuclide inventory for irradiated graphite-chlorine-36 inventory determination, in: Proceedings of IAEA Technical Committee Meeting held in Manchester, IAEA, Vienna, 1999, pp. 143-152.
[10]Roobottom, Thermochemical Radii of Complex Ions, Journal of Chemical Education, 76 (1999) 1570.
[11]L. Vieille, I. Rousselot, F. Leroux, J.-P. Besse, C. Taviot-Gueho, Hydrocalumite and Its Polymer Derivatives. 1. Reversible Thermal Behavior of Friedel's Salt: A Direct Observation by Means of High-Temperature in Situ Powder X-ray Diffraction, Chemistry of materials, 15 (2003) 4361-4368.
[12]G. Renaudin, J.P. Rapin, B. Humbert, M. François, Thermal behaviour of the nitrated AFm phase $Ca_4Al_2(OH)_{12}(NO_3)_2$ - $4H_2O$ and structure determination of the intermediate hydrate $Ca_4Al_2(OH)_{12}(NO_3)_2$ - $2H_2O$, Cement and Concrete Research, 30 (2000) 307-314.
[13]E. López-Salinas, M.E.L. Serrano, M.A.C. Jácome, I.S. Secora, Characterization of synthetic hydrocalumite-type $[Ca_2Al(OH)_6]NO_3{\cdot}mH_2O$: Effect of the calcination temperature, Journal of Porous Materials, 2 (1995) 291-297.
[14]J.D. Phillips, L.J. Vandeperre, Anion capture with calcium, aluminium and iron containing layered double hydroxides, Journal of Nuclear Materials, In Press, Corrected Proof.
[15]M.J. Hologado, V. Rives, S. San Román, Thermal decomposition of $Ca(OH)_2$ from acetylene manufacturing: a route to supports for methane oxidative coupling catalysts, Journal of Materials Science Letters, 11 (1992) 1708-1710.
[16]G. Fraissler, M. Jöller, T. Brunner, I. Obernberger, Influence of dry and humid gaseous atmosphere on the thermal decomposition of calcium chloride and its impact on the remove of heavy metals by chlorination, Chemical Engineering and Processing: Process Intensification, 48 (2009) 380-388.

BOTTOM-UP DESIGN OF A CEMENT FOR NUCLEAR WASTE ENCAPSULATION

T. Zhang,°,*, L.J. Vandeperre* and C. Cheeseman°
* Centre for Advanced Structural Ceramics and Department of Materials,
Imperial College London, South Kensington Campus, London SW7 2AZ, UK
° Department of Civil and Environmental Engineering, Imperial College London, South Kensington Campus, London, SW7 2AZ

ABSTRACT

The chemical interactions between waste and the cement, in which it is encapsulated, are dominated by the chemistry of the pore water in the cement. Hence tailoring the interaction between waste and cement requires designing the chemistry of the cement matrix. In this paper, the chemical design of a new cement compatible with aluminium metal will be discussed. The starting point was the realization that hydration of magnesia to form brucite should yield a pH around 10 and that this is a favourable pH for aluminium passivation, certainly when compared with the much more alkaline environment induced by Portland cement. A range of fillers was studied and metakaolin and fused silica found to be compatible with the desired pH. Further work with fused silica showed that the brucite reacts with the fused silica to give magnesium-silicate-hydrate (M-S-H) gel and that addition of magnesium carbonate allows buffering the early pH. Encapsulation trials with aluminium showed that little or no corrosion occurred in the time-frame studied.

INTRODUCTION

Cements are widely used in construction[1] but have many applications in waste management also[2-8]. For the latter, a first function of the cement is to encapsulate the waste in a low permeability matrix. A second function of the cement can be the immobilisation of the contaminants through either precipitation of the contaminants due the pH prevailing in the pore solution or because the contaminants are incorporated in phases that form upon cement hydration[8].

A key cement formulation for nuclear waste encapsulation in the UK is a blend of Portland cement (OPC) with blast furnace slag (BFS)[9, 10]. This system combines the remarkable ability of the hydration products of Portland cement to fill voids with a reduced heat of hydration by dilution with the much slower reacting BFS. The lower heat of hydration ensures that large volumes can be treated without cracking. It is recognized, however, that the availability of a range of cementitious binders to allow tailored matching of binder and waste, will enhance the ability to effectively treat all legacy nuclear waste[9]. Hence there has been a considerable effort in evaluating the potential of existing alternative cement formulations such as calcium aluminate cements, magnesium and calcium phosphate cements, calcium sulphoaluminate cements, alkali activated systems and geopolymers[9].

The idea explored here is that there is an advantage to designing cements for waste encapsulation from the bottom up, that is designing the chemistry of the cement to fit the application rather than what is often done to test whether existing cements, with sometimes complicated chemistries, would suit the application. In doing so it, the resulting cements should be simpler and the function of each component in the mortar composition can be understood. This should render making predictions of the long term performance more tractable. A disadvantage is that such novel cements do not have the same track record of successful applications in practice.

To illustrate the bottom-up approach, encapsulation of aluminium metal is used as a case study. For encapsulation of aluminium metal, the most important consideration is the potential continued corrosion of the metal in the cement matrix as it is well known that extensive reaction between the cement and aluminium occurs when aluminium is encapsulated in a blend of OPC with BSF[9]. Given that the corrosion of aluminium passivates for pH values in the 4 to 10 range[11], it was decided to develop a cement with a pore solution pH close to 10. This value was chosen because the solubility of many heavy metals is low around pH 10 so that this new cement could have wider applicability for the management of other waste.

Since the pH of the pore solution is determined largely by the presence of the more soluble phases in the assemblage of hydration products, it was decided to take advantage of a recently proposed alternative cement system based on the hydration of magnesium oxide (MgO)[12-16] to form magnesium hydroxide as a starting point for developing cements with a pH~10. Indeed, from the solubility of magnesium hydroxide (brucite), the pH of brucite in equilibrium with water is expected to be 10.5.

Therefore this paper records the different steps in the evolution of the cement formulation from pure MgO to a final blend consisting of MgO, magnesium carbonate ($MgCO_3$), silica fume (SF) and sand, and reports the results of encapsulation tests for aluminium metal.

EXPERIMENTAL

The materials and chemicals used to prepare the cement mixtures were a commercially available magnesium oxide (MgO, MagChem 30, M.A.F. Magnesite B.V., The Netherlands), silica fume (SF; Elkem Materials Ltd, UK), magnesium carbonate ($MgCO_3$, Fisher, UK) and metakaolin (MK, Metastar, UK). Additionally, sand obtained from Sibelco was made finer by milling to an average particle size of 50 μm. For comparison purposes some experiments were carried out with OPC (CEM-I, Lafarge UK) and with blends of OPC and BFS (GGBFS, Civil and Marine Slag Ltd., UK). The characteristics of the raw materials are summarised in Table 1.

Table 1. Characteristics of the raw materials as reported by the producer (PC = Portland cement, SF = silica fume, BFS = blast furnace slag, MK = Metakaolin).

	MgO	PC	SF	BFS	MK	MgCO3
Chemical composition (%)						
SiO_2	0.35	13.9	>97.5	34.7	59.5	<0.05
Al_2O_3	0.1	10.2	<0.7	13.9	34.0	-
Fe_2O_3	0.15	2.7	<0.3	0.5	0.7	<0.05
CaO	0.8	63.6	<0.3	38.8	0.6	<1
P_2O_5	-	-	<0.1	-	0.0	-
MgO	98.2	0.6	<0.5	9.2	0.5	-
K_2O	-	0.9	<0.6	0.28	2.0	<0.02
Na_2O	-	-	<0.3	0.25	0.0	<0.5
TiO_2	-	0.1	-	1.01	0.2	-
SO_3	0.05	6.9	<0.4	0.05	1.1	<0.5
Sum	99.65	98.9	-	98.7	99.6	-
LOI	1.7	-	<1.0	-	0.8	-
pH	-	-	8.2	-	4.8	10.2
Specific Gravity (g/cm^3)	3.23	3.15	1.94	2.15	2.14	-
Loose bulk density (kg/m^3)	350	-	311	174	285	-
Mean particle size (μm)	5	-	21.3	-	6.6	-
BET surface area (m^2/g)	25	0.4	21.4	-	17.3	-

The pH evolution with time and composition was studied by suspending 10 g of the cement mixture in 100 g of distilled water. The suspensions were continually agitated by rotation (10 rpm) except when measurements of the pH were taken, for which the solids were allowed to settle before determining the pH with a standard pH meter with a glass electrode. This method was selected for convenience after it was established that it gave the same results for the evolution of pH against time as when crushed fragments of cement mixtures were suspended in water after curing[17].

To interpret these measurements, a number of calculations of expected pH were made using Visual MINTEQ 2.6[18], a windows implementation of the MINTEQA2 algorithm for aquatic geochemical equilibria[19].

The hydration products were investigated using x-ray diffraction (XRD, Philips PW1720 powder diffractometer with a Cu Kα source) on crushed fragments of set cement blends cured in polyethylene bags.

The interaction of Al with selected binders was studied by encapsulating 6 x 3 x 25 mm strips of Al 1050 in the cement mixtures. In addition to simple observation of the interaction, the volume of H_2 generated from the samples through[9]

$$2Al + 2OH^- + 6H_2O \leftrightarrow 2[Al(OH)_4]^- + 3H_2(g) \quad (1)$$

was determined using the experimental set up shown in Figure 1. The latter experiments were conducted only for a reference composite cement (25:75 PC:BFS, w/s=0.33), a blend of 20:80 MgO:SF (w/s=1.6) and a more optimised sample from this work containing 20:5:25:50 $MgO:MgCO_3$:SF:Sand (w/s=0.35).

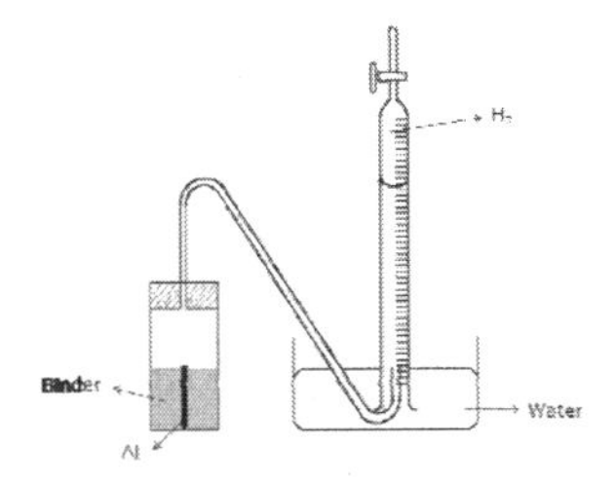

Figure 1. Experimental set-up to determine the hydrogen gas evolution versus time: an aluminium bar is buried partially in the cement in closed system which can measure the volume of gas evolved.

RESULTS AND DISCUSSION

Since pure MgO has a high heat of hydration, a range of experiments were conducted with potential filler materials: blast furnace slag (BFS), silica fume (SF) or metakaolin (MK). For initial tests the mixtures contained 50 wt% filler and 50 wt% MgO. As show in Figure 2, when using blast furnace slag, the pH is high and in fact comparable to what would be achieved with OPC. This is due to the high calcium oxide content of BFS. Blends with metakaolin show a more moderate pH, very similar to what was obtained when the MgO powder was suspended on its own. Over time a small reduction in pH is observed. In contrast, while blends with silica fume showed initially a pH close to that obtained with the MgO powder, in approximately 7 days a much lower pH is established, close to the aim of a pH of 10.

The pH of 11.3 when MgO is suspended in water is much higher than the expected value of 10.5 when brucite equilibrates with water. It is thought that the presence of small amounts of CaO in the MgO powder, see Table 1, causes the higher pH, and that the lowering of the pH with time when silica fume is added arises through the pozzolanic reaction which converts the calcium hydroxide resulting from the hydration of CaO into calcium silicate hydrate gel (CSH).

To investigate the potential pH values in blends with SF further, measurements of the pH after 28 days for a range of MgO and silica fume mixtures were made, see Figure 3. The pH appears to change gradually with composition, but on closer inspection it transpires that for blends containing more than 60 wt% MgO, the 28 days pH tends to 10.5, again consistent with pH control by brucite, whereas for mixtures with less than 30 wt% brucite, the pH equilibrates at about 9.5.

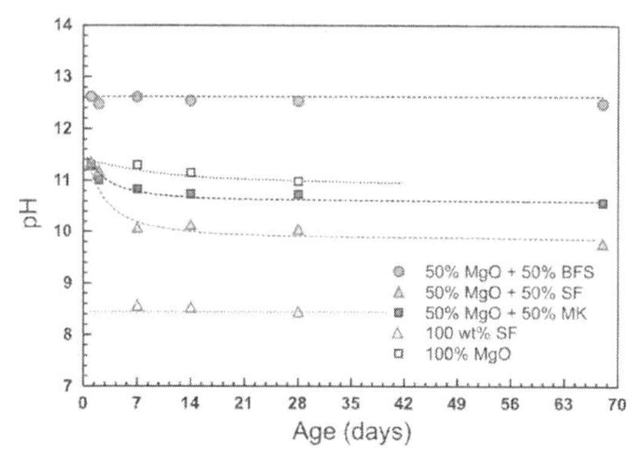

Figure 2. pH evolution with age for pure MgO, pure silica fume (SF) and blends of 50 wt% MgO and 50wt% blast furnace slag (BFS), silica fume (SF) or metakaolin (MK) as measured at an initial solids loading of 10 g per 100 g water.

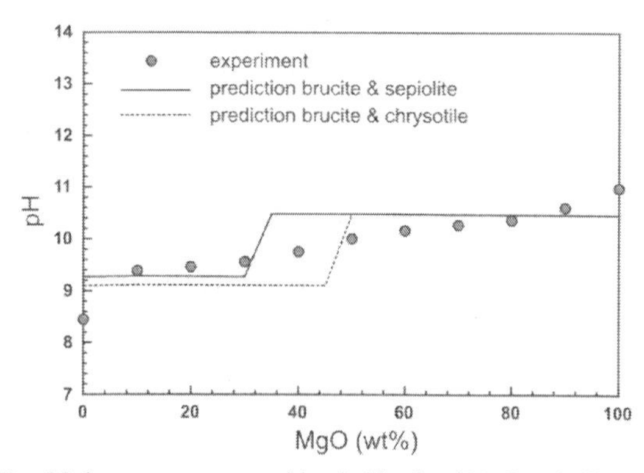

Figure 3. pH at 28 days versus composition in blends of MgO and silica fume compared with predictions for the pH assuming brucite and sepiolite can form or assuming brucite and chrysotile can form.

X-ray diffraction, see Figure 4, revealed that for 50 wt% MgO or higher in the blend, brucite is found as well as a poorly crystalline product as evidenced by the appearance of 2 additional broad peaks at 2θ ~35° and ~60° in addition to the broad peak due to silica fume at 2θ ~22°, whereas for blends with less than 50 wt% MgO, only the poorly crystalline hydration product remains.

The broad amorphous peaks were found to be consistent with the X-ray diffraction pattern of magnesium-silicate-hydrate gel (MSH) published by Brew and Glasser[20]. Hence, the change-over in the pH values observed as a function of composition appears to be due to the disappearance of brucite in mixtures with less than 50 wt% MgO by formation of MSH gel. To confirm this further, the pH versus composition was calculated whereby MSH was approximated either by sepiolite or chrysotile, two crystalline magnesium silicates with a composition close to MSH gel[20]. As shown in Figure 3, the calculations confirm that for low MgO contents, the pH should be close to 9-9.2, no brucite should remain and a pure magnesium silicate should precipitate. The higher solubility of the poorly crystalline material explains why the pH remains slightly higher than predicted, and the fact that the composition of the MSH gel is not as fixed as those of the crystalline products explains why the change in pH with composition is more gradual than predicted. At higher MgO contents, brucite is predicted to remain in the final phase assemblages and hence the pH should be about 10.5 as observed. Hence, although not intended, the silica fume reacts with the MgO forming a magnesium silicate hydrate gel. For excess MgO contents, the remainder of the MgO forms brucite. Therefore, the pH can be fine-tuned to just below 10 by limiting the MgO content so that after 28 days only MSH gel will present or to just above 10 by using more MgO.

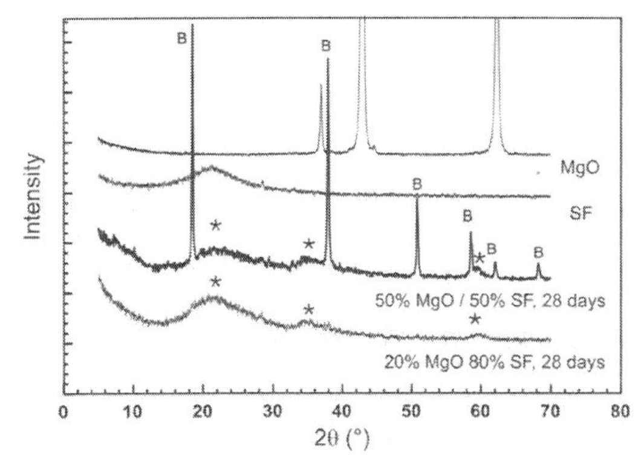

Figure 4. X-ray diffraction data for the MgO and silica fume (SF) starting materials and for two blends of MgO and silica fume after 28 days.

A first idea of the extent of reaction between aluminium and the binders was obtained by encapsulating strips of aluminium in the binder. As shown in Figure 5a, the reaction of aluminium with OPC CEM-I is so virulent that enough hydrogen gas forms to push the binder permanently away from the aluminium surface.

The reaction rate is reduced when the aluminium is encapsulated in a blend of OPC and BFS, see Figure 5b, but enough gas evolved to initially cover the aluminium strip in cement paste followed by the creation of pore channels by continued hydrogen gas evolution. More details on the reaction between aluminium and this binder can be found in the work of Milestone[9]. The reaction rate is reduced further in mixtures of MgO and silica fume, with a blend with only 20 wt% MgO showing less reaction than one with 50 wt% MgO. Quantification of some of these observations was obtained by measuring the amounts of hydrogen gas evolved, see Figure 6, and these confirmed that the hydrogen evolution rate was lower in a blend with 20 wt% MgO and 80 wt% SF compared to the reference CEM-I/BFS binder. However, it was clear also that substantial amounts of hydrogen still evolve from this binder. This is believed to be a consequence of the fact that in the first week, the pH is markedly higher, see Figure 2, than the target value.

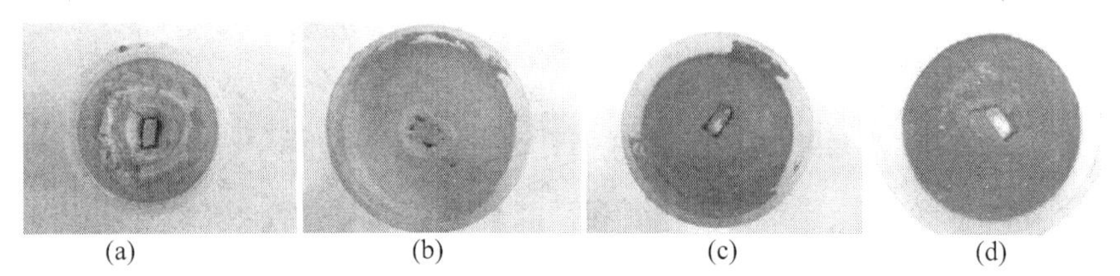

(a) (b) (c) (d)

Figure 5. Photographs taken 24 h after sample preparation showing the reaction between Al-1050 and a range of binders : (a) pure OPC CEM I, (b) 25 wt% OPC CEMI I and 75 wt% BSF, (c) 50 wt% MgO & 50 wt% silica fume and (d) 20 wt% MgO and 80 wt% silica fume. At the start of the experiment the amount of cement binder added left the Al strip visible as in (d). The diameter of the vessels is 37 mm.

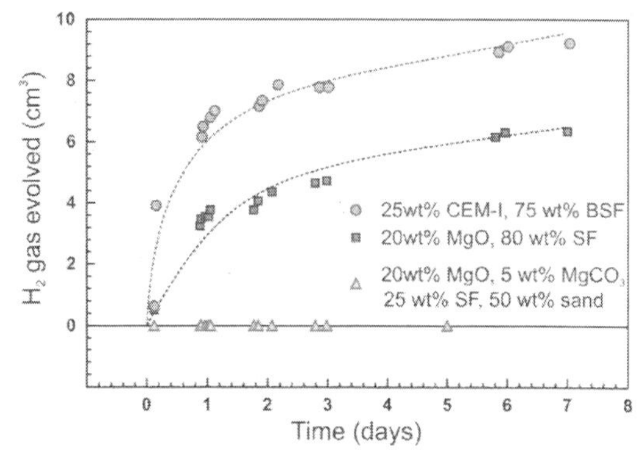

Figure 6. Hydrogen evolution against time for the reference blend of 25 wt% CEM-I and 75 wt% BFS, for 20 wt% MgO and 80 wt% SF and for 20 wt% MgO, 5 wt% $MgCO_3$, 25 wt% SF and 50 wt% sand.

Hence, to reduce the early pH, a cement component capable of buffering the pH in the early stages of the reaction was needed. Calculations of the total dissolved Mg concentration in equilibrium with a number of phases as a function of pH, see Figure 7, revealed that at high pH, magnesium carbonate ($MgCO_3$) is more soluble than magnesium hydroxide. Hence, in absence of a soluble silica source, the following reactions can be expected for pH values in excess of 10.2, i.e. for pH values in excess of the intersection point between the magnesium hydroxide and magnesium carbonate line (point a in Figure 7):

$$MgCO_3 \rightarrow Mg^{2+} + CO_3^{2-} \tag{2}$$

$$Mg^{2+} + 2OH^- \rightarrow Mg(OH)_2 \tag{3}$$

which reduces the pH. Conversely, when the pH becomes lower than 10.2, the solubility of brucite increases. Its dissolution according to

$$Mg(OH)_2 \rightarrow Mg^{2+} + 2OH^- \tag{4}$$

will then increase the pH again. Hence, addition of magnesium carbonate should buffer the pH. In the presence of soluble silica, hydrated magnesium silicates will form. In the presence of excess silica, the pH at which the system buffers could potentially be lowered to point b or b'. However, this lower pH value will only be reached once all brucite has been consumed and in the early stages brucite is always present because the hydration of the MgO powder is rapid relative to the dissolution of Si from silica fume. Therefore, reaching a lower pH value of about 8 due to equilibration at point b or b' is not expected. In agreement with these calculations, the early pH was reduced to 10.7 when some of the MgO was replaced by $MgCO_3$ in the cement formulation (35wt% MgO, 5 wt% $MgCO_3$, 60 wt% silica fume).

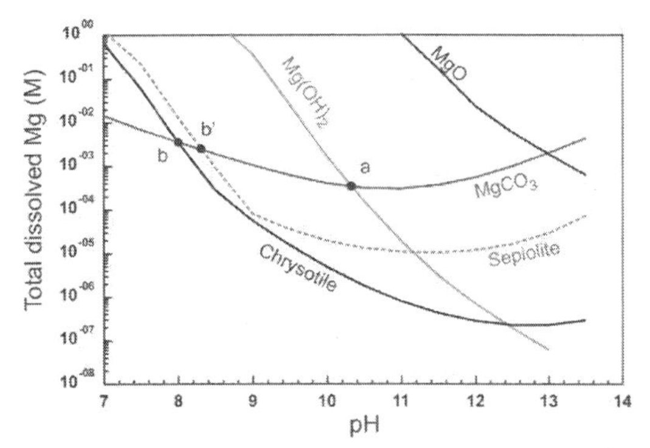

Figure 7. The total magnesium concentration in solution as a function of pH for equilibrium with a range of phases that can be present in the cement during hydration.

In a final step, sand was added to the formulation as well, effectively turning the cement into a mortar. This has the advantage of reducing the cost of the product as well as improving the dimensional stability by diluting the shrinkage strain when the gel dries[21, 22]. Adding sand reduces the buffer capacity of the cement as less of the pH controlling phases brucite and M-S-H gel will be present per unit volume and hence might not be appropriate if acid leaching is expected. Moreover, for compositions with excess brucite, quartz and brucite might react to form more M-S-H leading to depletion of the residual brucite, which would lower the pH from 10.5 to 9.5.

Figure 5 shows that when aluminium was buried in this final optimised binder containing magnesium carbonate to reduce the initial pH and sand no hydrogen gas was found to evolve, suggesting that the assumption that the corrosion was due to the high early pH in MgO-based blends was correct.

CONCLUSIONS

A novel cement system with a pH around 10 was designed by selecting the desired hydrated phase, brucite. In the presence of silica fume, added originally as a filler to reduce the heat of hydration and as a pozollan to react with calcium impurities, a poorly crystalline hydrated magnesium silicate gel forms. The prevailing pH is either just below 10 or just above 10 depending on the relative MgO to silica fume ratio. However, these pH values are only reached after 7 to 14 days, and the higher initial pH value causes substantial corrosion when aluminium is encapsulated in the cement. This can be overcome by addition of magnesium carbonate, which buffers the early pH to around 10.2. When aluminium was encapsulated in this optimised cement, there was no measurable reaction in the timeframe studied.

ACKNOWLEDGEMENTS

The financial support from the Engineering and Physical Sciences Council (EPSRC) of the United Kingdom through the University consortium for decommissioning and management of nuclear waste is gratefully acknowledged.

REFERENCES

1. S. Mindess, J.F. Young, and D. Darwin, Concrete. second ed, Upper Saddle River: Prentice Hall. 644 (2003).
2. M. Yousuf, A. Mollah, R.K. Vempati, T.C. Lin, and D.L. Cocke, The interfacial chemistry of solidification/stabilization of metals in cement and pozzolanic material systems. *Waste Management*, **15**(2): p. 137-148 (1995).
3. R.D. Spence, Stabilization of high salt waste using a cementitious process, in *Innovative Technology summary report*. Department of Energy, USA (1999).
4. J. Paya, M.V. Borrachero, J. Monzo, and M. Bonilla, Properties of Portland cement mortars incorporating high amounts of oil-fuel ashes. *Waste Management*, **19**: p. 1-7 (1999).
5. C.M. Jantzen, F.P. Glasser, and E.E. Lachowski, Radioactive waste-Portland cement systems: I, Radionuclide distribution. *Journal of the American Ceramic Society*, **67**(10): p. 668-673 (1984).
6. S. Asavapisit, G. Fowler, and C.R. Cheeseman, Solution chemistry during cement hydration in the presence of metal hydroxide wastes. *Cement and Concrete Research*, **27**(8): p. 1249-1260 (1997).
7. J.R. Fitch and C.R. Cheeseman, Characterisation of environmentally exposed cement-based stabilised/solidified industrial waste. *Journal of Hazardous Materials*, **101**(3): p. 239-255 (2003).
8. B. Batchelor, Overview of waste stabilization with cement. *Waste management*, **26**(7): p. 689-698 (2006).
9. N.B. Milestone, Reactions in cement encapsulated nuclear wastes: need for toolbox of different cement types. *Advances in Applied Ceramics*, **105**(1): p. 13-20 (2006).
10. R.J. Caldwell, S. Rawlinson, E.J. Butcher, and I.H. Godfrey. Characterisation of full-scale historic inactive cement-based intermediate level wasteforms. in *Stabilisation/Solidification Treatment and Remediation*. Cambridge: A.A. Balkema Publishers (2005).
11. E. Deltombe, C. Vanleugenhaghe, and M. Pourbaix, Aluminium. *Atlas of Electrochemical Equilibria in Aqueous Solutions*: p. 168-175
12. L. Vandeperre, M. Liska, and A. Al-Tabbaa. Reactive MgO cements : Properties and Applications. in *International conference on sustainable construction materials and technologies*. Coventry: Taylor and Francis (2007).
13. L.J. Vandeperre and A. Al-Tabbaa, Accelerated carbonation of reactive MgO cements. *Advances in Cement Research*, **19**(2): p. 67-79 (2007).

14. M. Liska, L.J. Vandeperre, and A. Al-Tabbaa, Influence of carbonation on the properties of reactive magnesia cement-based pressed masonry units. *advances in Cement Research*, ***20***(2): p. 53-64 (2008).
15. L.J. Vandeperre, M. Liska, and A. Al-Tabbaa, Microstructures of reactive magnesia cement blends. *Cement & Concrete Composites*, **30**: p. 706-716 (2008).
16. L.J. Vandeperre, M. Liska, and A. Al-Tabbaa, Hydration and Mechanical Properties of Magnesia, Pulverized Fuel Ash, and Portland Cement Blends. *Journal of Materials in Civil Engineering*, **20**: p. 375-383 (2008).
17. T. Zhang, C.R. Cheeseman, and L.J. Vandeperre, Development of novel low pH cement systems for encapsulation of wastes containing aluminium, in *Decommissioning, Immobilisation and Management of Nuclear Waste for Disposal, Diamond '09 Conference*: York (2009).
18. J.P. Gustafsson, Visual MINTEQ. KTH, Dept. of Land and Water Resources Engineering: Stockholm (2009).
19. J.D. Allison, D.S. Brown, and K.J. Novo-Gradac, MINTEQA2/PRODEFA2 A geochemical assessment model for environmental systems. U.S. Environmental Protection Agency (1991).
20. D.R.M. Brew and F.P. Glasser, Synthesis and characterisation of magnesium silicate hydrate gels. *Cement and Concrete Research*, **35**: p. 85-98 (2005).
21. M.D. Cohen and B. Mobasher, Drying shrinkage of expansive cements. *Journal of Materials Science*, **23**: p. 1976-1980 (1988).
22. P. Grassl, H. Wong, and N. Buenfeld, Influence of aggregate size and volume fraction on shrinkage induced micro-cracking of concrete and mortar. *Cement and concrete research*, **40**(1): p. 85-93 (2010).

Fuel Ceramics and Irradiation Effects

MICROSTRUCTURAL ANALYSIS OF SECONDARY PHASES IN SILICON CARBIDE FABRICATED WITH SIC NANO-POWDER AND SINTERING ADDITIVES

Takaaki Koyanagi
Graduate School of Energy Science, Kyoto University
Gokasho Uji, Kyoto 611-0011, Japan

Sosuke Kondo and Tatsuya Hinoki
Institute of Advanced Energy, Kyoto University
Graduate School of Energy Science, Kyoto University
Gokasho Uji, Kyoto 611-0011, Japan

ABSTRACT

Chemical composition, structure and distribution of secondary phases in two kinds of monolithic SiC ceramics fabricated by liquid phase sintering with SiC nano-powder and 6 or 9 wt.% sintering additives of Al_2O_3 and Y_2O_3 were investigated. The major secondary phase in both samples was identified as crystalline yttrium aluminum garnet (YAG) by XRD pattern. EPMA-WDS analysis revealed that the sample containing 9 wt.% additives had the relatively distinct Y-rich and Al-rich regions, by contrast with relatively-homogeneous distribution of the additives elements in the sample containing 6 wt.% additives. The dominating secondary phases at triple junctions of YAG for Y-rich regions and Al_2O_3 for Al-rich regions in SiC (9 wt.% additives) were identified by TEM-EELS analysis. In SiC (6 wt.% additives), the dominating YAG was observed at triple junctions. High-resolution TEM revealed that amorphous films with content Y, Al and O existed at both the SiC-SiC and SiC-YAG grain boundaries.

INTRODUCTION

Silicon carbide (SiC) fiber reinforced SiC matrix composites have been developed for nuclear fusion and fission energy systems because of the inherent high-temperature capability and low activation characteristics [1, 2]. One of the fabrication process for SiC/SiC composites is a nano-infiltration and transient eutectic-phase (NITE) process, which was developed making use of the liquid phase sintering (LPS) with SiC nano-powder and small amounts of oxide additives [3]. It was reported that the dense NITE SiC/SiC composites surpassed the chemical vapor infiltrated (CVI) SiC/SiC composites in matrix cracking stress, thermal conductivity, and gas permeability [3-5]. Furthermore, the tensile fracture behavior is easily controllable by tailoring the amounts of additives and fabrication temperature for NITE SiC/SiC composites, and two types of composites showing high PLS or high ductility are now available [6]. Thus, NITE composites have been selected as one of candidates for fusion reactor components in the ITER broader approach (BA) activities in an attempt to improve the plant-efficiency [1]. In the BA activities, irradiation effects on monolithic SiC samples representing matrix of NITE SiC/SiC composites have been studied to accumulate the irradiation data regarding the SiC containing secondary phases and to compare with the data from high purity chemical vapor deposited (CVD) SiC in the early stage of the BA activities.

Because the microstructural change by irradiation, especially in non-metallic materials, is known to be primary cause for volume expansion (swelling) and change in thermal conductivity, microstructural stability under irradiation has been extensively studied for high purity CVD SiC [7-9]. Because somewhat different microstructural evolution may be anticipated in SiC fabricated with sintering additives due to the distribution of the secondary phases, the microstructural information is necessary to understand the difference in the irradiation tolerances of CVD SiC. It has been reported that impurity or secondary phase in SiC affected the magnitude of and the dose dependency of the swelling [10], where the amorphization of the secondary phase by ion irradiation increased the total amount of the SiC swelling [11]. It was also reported that the reduction of irradiated strength of reaction bonded SiC containing excess Si, which was attributed to internal crack caused by differential swelling between SiC and Si [12]. However, the knowledge of the secondary phases in SiC is extremely limited even for the most basic information such as the unirradiated chemical compositions and the irradiation effect. The objective in this work is to identify the composition, structure and distribution of the secondary phase in two types of monolithic SiC representing matrix of the high PLS and high ductility NITE SiC/SiC composites. Potential irradiation effects will be discussed based on the differences from the high purity CVD SiC.

EXPERIMENTAL PROCEDURE

Samples used were made of β-SiC nano-powder, and Al_2O_3 and Y_2O_3 sintering additives. A total of 6 or 9 wt.% additives were mixed with SiC powders and then sintered by hot-pressing. These samples were supplied by Japan Atomic Energy Agency through BA activities. In this paper, two types of monolithic SiC ceramics referred to as AD6 containing 6 wt.% additives and AD9 containing 9 wt.% additives were studied. The AD6 represents the matrix of high PLS NITE composite, and AD9 represents the matrix of high ductility NITE composite, respectively.

The compositional distribution profiles of samples were investigated using a JEOL JXA-8500F electron probe micro analyzer (EPMA) at 15 kV equipped with wavelength dispersive spectrometers (WDS). Standard materials used for the quantitative WDS point analysis included CVD SiC produced by Rohm and Haas Advanced Materials and $Y_3Al_5O_{12}$ (yttrium aluminum garnet, YAG) produced by Kojundo Chemical Industries. In the WDS panalysis, more than 10 points were tested for each sample. The secondary phase in samples was confirmed by X-ray diffraction (XRD) pattern. The microstructural examination was performed using a JEOL JEM-2200FS field-emission transmission electron microscope (FE-TEM) operated at 200 kV. The elemental analysis was performed by the scanning-transmission electron microscope equipped with a energy dispersive spectroscope (STEM-EDS). The chemical composition or structure of the secondary phase was identified by the electron energy-loss spectroscope (EELS). The grain boundaries in samples were analyzed by the high-resolution electron microscope (HREM).

RESULTS AND DISCUSSION

The XRD patterns of AD6 and AD9 show that the both specimens contained YAG phase beside the main β-SiC phase as illustrated in Fig. 1, indicating that YAG was the main crystalized secondary phase in both materials. The existence of YAG is consistent with yttria and alumina phase diagram, and reported in LPS SiC fabricated with those additives [13, 14].

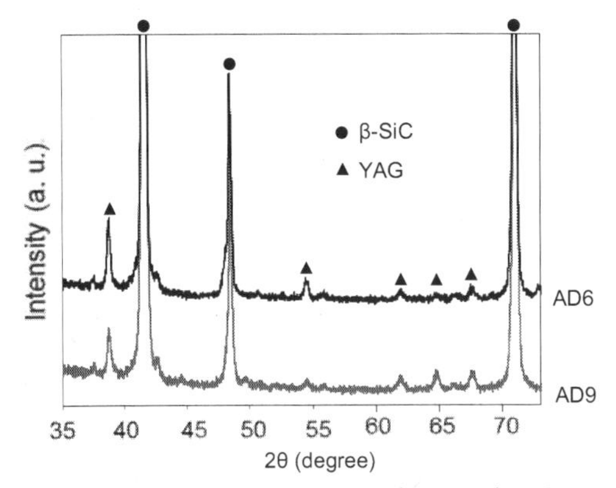

Figure 1. The XRD patterns of AD6 and AD9.

The elemental maps by WDS analysis show the distribution of the additives in AD6 (Fig. 2 (a)) and AD9 (Fig. 2 (b)). It is noted that the region in AD9 was composed of two different phases, where the one showed relative Y-rich and the other showed relative Al-rich composition. The Al-rich region accounted for approximately 40% of the surface area of AD9, and one Al-rich area surrounded by Y-rich region was ~5×10^{-4} mm^2. By contrast, in AD6, relatively-homogeneous

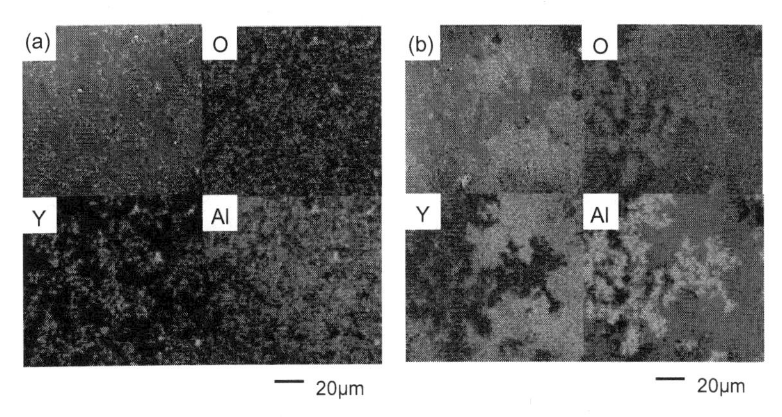

Figure 2. WDS elemental maps illustrating distribution patterns of the additive elements in (a) AD6 and (b) AD9.

distributions of the elements were observed. Table 1 shows the results of the quantitative WDS analysis for AD6 and AD9. The most significant difference of the elemental composition between Y-rich and Al-rich regions was the amount of yttrium.

Figure 3 shows the typical STEM images and EDS elemental maps for (a) AD6, (b) Y-rich and (c) Al-rich regions in AD9. The locations, in which the additives elements were observed, are

Table 1 Secondary phases, the location of them and chemical composition in 6AD and 9AD.

Sample	Secondary Phases	Locations of secondary phases	Y content (atm.%)	Al content (atm.%)	O content (atm.%)	Total (atm.%)
AD6	YAG Al_2O_3[a] Amorphous film	TJ[c] TJ GB[d]	0.42	1.5	3.1	5.0
Y-rich region in AD9	YAG Amorphous film	TJ GB	1.2	2.4	5.9	9.5
Al-rich region in AD9	Al_2O_3 Y-Al oxide[a,b] Amorphous film	TJ TJ GB	0.16	2.6	5.3	8.1

a Minor secondary phase
b Not identified
c Triple junction
d Grain boundary

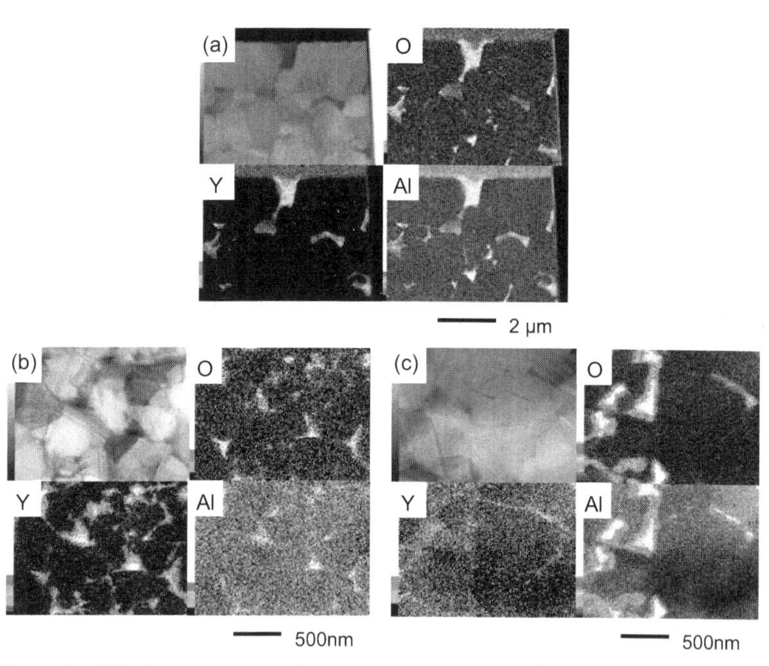

Figure 3. STEM images and EDS elemental maps illustrating distribution patterns of the additive elements in (a) AD6, (b) Y-rich region and (c) Al-rich region of AD9.

summarized in Table 1. Most of the secondary phase (~90% of the total secondary phase area) was observed at triple junctions, and the rest was uniformly distributed along grain boundaries in all samples. In AD9, the dominating secondary phases observed were Y-Al oxides for Y-rich regions and the Al oxides in addition to the minor Y-Al oxides for Al-rich regions. In AD6, the dominating Y-Al oxides were observed with the minor Al oxides. A difference in the grain size between AD9 and AD6 was also observed, where the average grain sizes of SiC were 830 nm in AD6 and 260 nm in AD9. No significant deference in the grain size between Y-rich and Al-rich regions were measured in AD9. No elements related to the additives were detected inside SiC grains for all samples.

The Y-Al and Al oxides at triple junctions in both AD6 and AD9 were identified by EELS technique, and the results are listed in Table 1. Figure 4 shows the comparison of the typical EELS spectrums of Al-$L_{2,3}$ and O-K edges between the Y-Al and Al oxides in SiC samples and bulk YAG. The similar shape of EELS spectrums to bulk YAG identified the Y-Al oxides in SiC samples as YAG. More than 90% area ratio of Y-Al oxides in AD6 and Y-rich region of AD9 were distinguished as YAG. This result is consistent with XRD results in Fig. 1. The EELS spectrum of Al oxides shows the existence of Al_2O_3 due to the similar shape of the reported alumina spectrum [15]. Other secondary phases in Al-rich region weren't identified because of small amount of the segregation. The combination of WDS, EDS and EELS analysis revealed that Y-rich and Al-rich regions in AD9 were the regions where YAG or Al_2O_3 existed as a major secondary phase at triple junctions, respectively. In addition, in AD6, YAG was the major secondary phase at triple junctions.

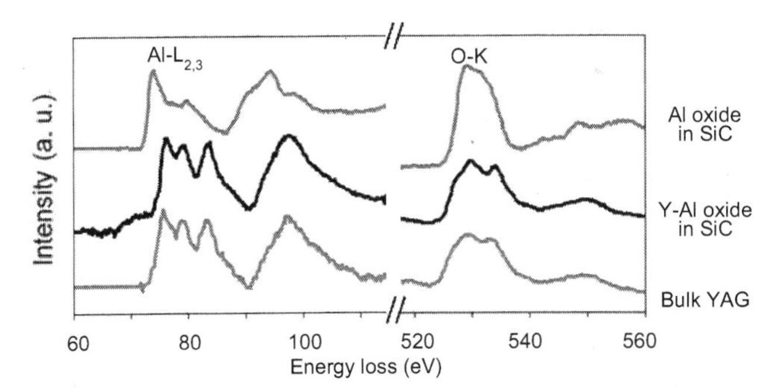

Figure 4. Comparison of the typical EELS spectrums of Al-$L_{2,3}$ and O-K edges between the Al and Y- Al oxides in SiC samples and bulk YAG.

Figure 5 shows the HREM images of (a) SiC-YAG and (b) SiC-SiC grain boundaries at Y-rich region in AD9, revealing the existence of the amorphous intergranular films with ~2 nm thickness, and content of Y, Al and O based on the EDS point analysis. The amorphous film between grain boundaries described above was also observed in both AD6 and Al-rich region in AD9. In the all samples, amorphous films were essentially found along grain boundaries. The existence of intergranular films can be explained by the result of two competing interactions, one is an attractive van der Waals dispersion force between the grains, and the other is a structural disjoining repulsive force between the grain and amorphous phase [16]. The amorphous film is known to affect creep behavior of ceramics as discussed below. Shimoda et. al. reported that one of the thermal creep mechanisms of SiC made of SiC nano-powder and oxide additives was the viscos flow of the amorphous intergranular film which is absent from the high purity CVD SiC with clean grain boundaries [17]. Viscosity of the film potentially modified under irradiation as reported in that of vitreous silica under proton irradiation at ~ 600 °C corresponding to thermal viscosity at 1200 °C [18]. This implies that the irradiation induced creep of SiC containing amorphous films is possible by the viscous flow mechanism at the temperature in which contribution of thermal creep is insignificant.

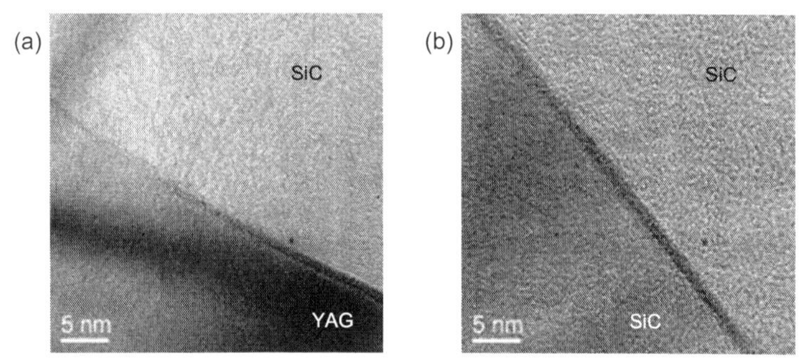

Figure 5. HREM images revealing the presence of an amorphous intergranular film located at (a) YAG-SiC and (b) SiC-SiC grain boundaries in AD9.

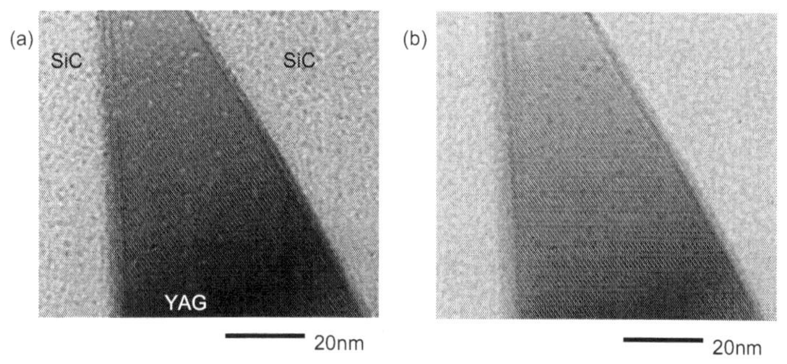

Figure 6. The cavities in the YAG phase in AD9 under imaging conditions of (a) under focus and (b) over focus.

The cavity formation at YAG phase in AD9 was shown in Fig. 6. The cavity images are taken under the conditions of (a) under focus and (b) over focus to show the contrast change typically used to verify the presence of cavities. The cavities were also detected at YAG in AD6. The average cavity sizes were 1.2 and 1.5 nm in AD6 and AD9, respectively. J. Ihle et. al. reported the chemical interaction in Al_2O_3-Y_2O_3-SiC system [19]. According to their thermodynamic calculations, the possible reactant gases during the processing in this study are CO, SiO, Al_2O and Al gases. Therefore, these spherical cavities observed in YAG phase in Fig. 6 (a) and (b) are likely stabilized by the pressure associated with the phase gases.

In the previous ion irradiation experiment, similarity of the 0.5 ~ 0.6% swelling between AD6, AD9 and CVD SiC was reported at 1000 °C to 3dpa, indicating that the swelling of secondary phases wasn't significantly large in comparison with that of CVD SiC [11]. This result is consistent with comparable 0.5% swelling between CVD SiC and bulk YAG at 1000 °C to 2 dpa-SiC [7, 20], because the major secondary phase in both AD6 and AD9 was identified as YAG in this study. It was also reported that the elastic moduli of AD6 and AD9 didn't significantly change by irradiation at 1000 °C [11]. One of the reasons for maintaining the modulus is the similar swelling between SiC and YAG (secondary phase), because if the differential swelling between them introduced significant misfit strain, consequent crack generation should degrade the modulus as reported in the case of reaction bonded SiC with excess Si [12].

The amorphized Y-Al oxides at Y-rich regions in AD9 were observed following ion irradiation at 280 °C where amorphization of SiC grains did not occur [11]. These Y-Al oxides were mostly identified as YAG in this work. The irradiation induced amorphization was a well-known phenomenon for materials under ion irradiation. H. M. Naguib et. al. proposed that amorphization by ion irradiation occurs whenever the ratio (crystallization temperature)/(melting point) exceeds 0.30 [21]. The irradiation induced amorphization of YAG was consistent with this criterion due to the reported crystallization temperature of 1100~ °C [22] and melting point of 1940 °C [23].

SUMMARY

Chemical composition, structure and distribution of secondary phase in the monolithic SiC ceramics containing 6 or 9 wt.% sintering additives of Y_2O_3 and Al_2O_3 were investigated. The sample containing 9 wt.% additives was divided into the relatively Y-rich and Al-rich regions, by contrast with relatively-homogeneous distribution of the elements in the SiC containing 6 wt.% additives. Most of additives were existed at triple junctions in both materials. In the SiC (9 wt.% additives), the dominating secondary phases observed at triple junctions were YAG for Y-rich regions, and the Al_2O_3 in addition to the minor Y-Al oxides for Al-rich regions. In the SiC (6 wt.% additives), the dominating YAG were observed with the minor Al_2O_3 at triple junctions. The amorphous intergranular films with content of Y, Al and O existed at both the SiC-SiC and SiC-YAG grain boundaries in both samples.

ACKNOWLEDGEMENT

The authors wish to thank Japan Atomic Energy Agency for supplying the SiC ceramics. This work was partially supported by The Broader Approach Activities in the Field of Fusion Energy Research.

REFERENCES

[1] T. Nozawa, T. Hinoki, A. Hasegawa, A. Kohyama, Y. Katoh, L. L. Snead, C.H.

Henager Jr. and J.B.J. Hegeman, Recent advances and issues in development of silicon carbide composites for fusion applications, *J. Nucl. Mater.*, **386-8**, 622-7 (2009).
[2] A. Kohyama, T. Hinoki, T. Mizuno, M. Sato, Y. Katoh, J. S. Park, R & D of Advanced Material Systems for Reactor Core Component of Gas Cooled Fast Reactor, *Proc. Amer. Nucl. Soc.*, **3**, 1451-60 (2005).
[3] Y. Katoh, S. M. Dong, A. Kohyama, Thermo-mechanical properties and microstructure of silicon carbide composites fabricated by nano-infiltrated transient eutectoid process, *Fusion Eng. Des.*, **61-2**, 723-31 (2002).
[4] Y. Katoh, A. Kohyama, T. Nozawa and M. Sato, SiC/SiC composites through transient eutectic-phase route for fusion applications, *J. Nucl. Mater.*, *329-333*, 587-91 (2004).
[5] T. Hino, E. Hayashishita, Y. Yamauchi, M. Hashiba, Y. Hirohata and A. Kohyama, Helium gas permeability of SiC/SiC composite used for in-vessel components of nuclear fusion reactor, *Fusion Eng. Des.*, **73**, 51-6 (2005).
[6] K. Shimoda, J.S. Park, T. Hinoki and A. Kohyama, Microstructural optimization of high-temperature SiC/SiC composites by NITE process, *J. Nucl. Mater.*, **386-8**, 634-8 (2009).
[7] L. L. Snead, T. Nozawa, Y. Katoh, T. S. Byun, S. Kondo, D. A. Petti, Handbook of SiC properties for fuel performance modeling, *J. Nucl. Mater.*, **371**, 329-77 (2007).
[8] G. E. Youngblood, D. J. Senor and R.H. Jones, Effects of irradiation and post-irradiation annealing on the thermal conductivity/diffusivity of monolithic SiC and f-SiC/SiC composites, *J. Nucl. Mater.*, **329-33**, 507-12 (2004).
[9] S. Kondo, Y. Katoh and L. L. Snead, Microstructural defects in SiC neutron irradiated at very high temperatures, *J. Nucl. Mater.*, **382**, 160-9 (2008).
[10] H. Huang and N. Ghoniem, A swelling model for stoichiometric SiC at temperatures below 1000°C under neutron irradiation, *J. Nucl. Mater.*, **250**, 192-9 (1997).
[11] T. Koyanagi, S. Kondo and T. Hinoki, The influence of sintering additives on the irradiation resistance of NITE SiC, *J. Nucl. Mater.*, doi:10.1016/j.jnucmat.2010.12.181.
[12] R.B. MATTHEWS, IRRADIATION DAMAGE IN REACTION-BONDED SILICON CARBIDE, *J. Nucl. Mater.*, **51**, 203-8 (1974).
[13] O. Fabrichnaya, H. J. Seifert, T. Ludwig, F. Aldinger and A. Navrotsky, The assessment of thermodynamic parameters in the Al_2O_3-Y_2O_3 system and phase relations in the Y-Al-O system, *Scand. J. Metal.*, **30**, 175-183 (2001).
[14] R. Huang, H. Gu, J. Zhang and D. Jiang, Effect of Y_2O_3-Al_2O_3 ratio on inter-granular phases and films in tape-casting α-SiC with high toughness, Acta. Mater., **53**, 2521-9 (2005).
[15] M. A. Gulugun, W. Y. Ching and M. Ruhle, Yttrium-Segregated Grain Boundaries in α- Al_2O_3: An EELS Study, *Mater. Sci. Forum*, **294-6**, 289-92 (1999).
[16] D. R. Clarke, On the Equilibrium Thickness of Intergranular Glass Phases in Ceramic Materials, *J. Am. Ceram. Soc.*, **70**, 15-22 (1987).
[17] K. Shimoda, S. Kondo, T. Hinoki and A. Kohyama, Thermal stress relaxation creep and microstructural evolutions of nanostructured SiC ceramics by liquid phase sintering, *J. Eur. Ceram. Soc.*, **30**, 2643-52 (2010).
[18] Z. Zhu and P. Jung, Irradiation induced dimensional changes in ceramics, *Nucl. Instr. and Meth. in Phys. Res. B*, **91**, 269-73 (1994).
[19] J. Ihle, M. Herrmann and J. Adler, Phase formation in porous liquid phase sintered silicon carbide: Part III: Interaction between Al_2O_3–Y_2O_3 and SiC, *J. Eur. Ceram. Soc.*, **25**, 1005-13 (2005).
[20] R. J. M. Konings , K. Bakker, J.G. Boshoven, R. Conrad and H. Hein, The influence of neutron irradiation on the microstructure of Al_2O_3, $MgAl_2O_4$, $Y_3Al_5O_{12}$ and CeO_2, *J. Nucl. Mater.*, **254**, 135-42 (1998).
[21] H. M. Naguib and R. Kelly, Criteria for bombardment induced structural changes in non metallic solids, *Rad. Effects*, **25**, 1-12 (1975).
[22] J. G. Li, T. Ikegami, J. H. Lee, and T. Mori, Well-sinterable $Y_3Al_5O_{12}$ powder from carbonate precursor, *J. Mater. Res.*, **15** (7), 1514-1523 (2000).

[23] J. L. Caslavsky and D. J. Viechnicki, Melting behaviour and metastability of Y aluminium garnet (YAG) and $YAlO_3$ determined by optical differential thermal analysis, *J. Mater. Sci.*, **15**, 1709-1718 (1980).
[24] X. Kerbiriou, J. M. Costantini, M. Sauzay, S. Sorieul, L. Thomé, J. Jagielski, and J. J. Grob, Amorphization and dynamic annealing of hexagonal SiC upon heavy-ion irradiation: Effects on swelling and mechanical properties, *J. Appl. Phys.*, **105**, 073513 (2009).
[25] S. Rios, E. K. H. Salje, M. Zhang and R. C. Ewing, Amorphization in zircon: evidence for direct impact damage, *J. Phys.: Condens. Matter*, **12**, 2401–12 (2000).

MEASUREMENTS OF IRRADIATION CREEP STRAIN IN SILICON CARBIDE IRRADIATED WITH SILICON IONS

S. Kondo, T. Koyanagi, T. Hinoki, O. Hashitomi
Institute of Advanced Energy, Kyoto University
Gokasho Uji, Kyoto 611-0011, Japan

ABSTRACT

The significant internal share stress through the differential swelling will be introduced in the thick structures of nuclear reactors by the tans-thickness temperature gradient. In this work, the irradiation creep in SiC, which is indispensable to mitigate the internal stress, was studied by using ion-irradiation. High purity CVD SiC samples mechanically thinned to 50, 75, and 100 μm were firmly fixed on the curved irradiation base and the tensioned surfaces were irradiated to 0.01-3 dpa at 400, 600, and 800°C. The initial applied stresses at the surface were 150, 225, and 300 MPa, respectively, depending on the sample thickness. The creep strain was estimated from the sample curvature based on comparison with the curvature of sample irradiated under non-stressed condition. Creep strain were strongly dependent on both the applied stress and irradiation temperature, where the approximate linear relationship with swelling rates measured for non-stressed SiC was found for all the irradiation temperatures. Creep strains were also linearly dependent on the applied stress with temperature dependent proportionality constants.

INTRODUCTION

Silicon carbide (SiC) and SiC/SiC composites are promising materials for fuel coating and fusion blankets due to the exceptionally high tolerances of the material properties to the neutron irradiation even at high temperatures. Excellent irradiation stability of SiC at wide temperature range and damage levels in nuclear reactors has been reported so far [1]. It is well accepted that the radiation swelling becomes saturate at ~0.1 dpa and is negatively dependent on the irradiation temperature below 1000°C in addition to the retention of unirradiated strength [2, 3]. These results demonstrated that the feasibility of the functional use of SiC materials in a simplified irradiation condition. However, the significant internal share stress may be introduced when they are used as a thermal insulator such as flow-channel-insert wall through the differential swelling related to the through-thickness temperature gradient [4]. Irradiaition-induced creep of SiC, which is one of the key fundamental irradiation properties lacking in certain data, may be indispensable to mitigate the internal stress. Recent neutron data pointed out that the importance of understanding of the irradiation creep at early stage of the irradiation because of the larger creep rates comparing to that in steady-state creep regime [5]. In case such that the creep strain is sensitive to the damage levels, somewhat systematical irradiation experiments under well controlled irradiation conditions may be needed. Ion-irradiation methods may allow much quick and simple experiment under accurately controlled irradiation conditions. In this report, the irradiation-induced creep in SiC was studied by ion-irradiation method developed in this study.

EXPERIMENTAL PROCEDURE

The material used for this work was polycrystalline -SiC produced through chemical vapor deposition (CVD) by Rohm and Haas Advanced Materials (Woburn, Massachusetts, USA). The CVD material is very high purity, with typical total metal impurity concentration less than 5 wppm. The grain size is between 5 and 10 μm in the plane parallel to the deposition substrate, with the grains elongated in the <111> growth direction perpendicular to the substrate. The material is typically free of micro-cracks or other large flaws, but atomic layer stacking faults on the {111} planes are common. Density as measured using Archimedes' principle is essentially the theoretical density (3.210 g/cm^3).

Samples and fixtures for the ion-irradiation experiments are shown in Fig. 1 (a) and are schematically shown in Fig. 1 (b). Samples with dimensions of 1.5^w × 3.0^l mm were mechanically thinned and lap finished for both surfaces aiming to 50, 75, or 100 μm thicknesses. The machining errors of the thickness were less than ± 1.5 μm. Each sample was sandwiched between upper and lower fixtures in the order as schematically shown in Fig. 1 (b), and then samples were tightly fixed by outer holding fixture (the outer fixture is not shown in the figure) during annealing or ion-irradiation. The upper and lower fixtures have opposite sign of the curvature of 75 mm in radius. A rectangular slit with 5 mm in length and 0.5 mm in width was provided at the central location of each upper fixture. The tensioned surfaces of the curved samples were subjected to ion-beam for the cases of irradiation with applied stress as shown in Fig. 1. The same fixtures were also used for annealing of control specimens, where specimens were heated in the irradiation chamber for 2 or 3 hours at 400, 600, and 800°C, respectively. Other sets of straight fixtures with similar dimensions without the curvature were used for the cases of irradiation without the stress. Detailed descriptions of the fixtures developed may be found in elsewhere [6].

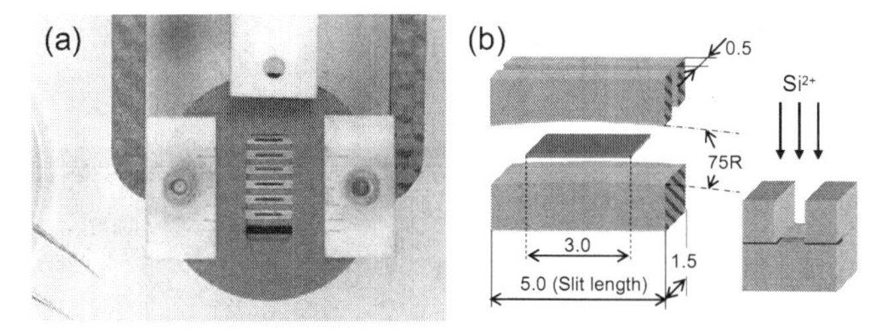

Figure 1. Samples and fixtures for the ion-irradiation experiments under applied stress. Overview of the irradiation holder including six sets of samples and fixtures are shown in (a). A -SiC bar is placed in the very bottom slot for the in-situ temperature monitoring. Si-ions passing through each slit with 1.5 nm width were irradiated to the tensioned sample surfaces as schematically shown in (b).

The samples were irradiated with 5.1 MeV Si^{2+}-ions at 400, 600, and 800°C up to nominally 3 dpa (displacement per atom) in DuET facility at Kyoto University [7]. Unstressed samples were also irradiated at 400-800°C up to 3 dpa in the straight fixtures. The depth profiles of displacement damage and concentration of implanted Si for the case of averaged damage level of 1 dpa were calculated by SRIM 98 and were shown in Fig. 2. The initial applied stresses at the irradiated surface were 150, 225, and 300 MPa, respectively, depending on the sample thickness. The surface curvatures in a stress axis, corresponding to the longitudinal direction of the thin strip samples, of annealed (unirradiated) and irradiated samples were recorded using atomic force microscopy (AFM) in tapping mode with the very compliant cantilever on a VN-8000 from KEYENCE. The curvature at the longitudinal centerline was confirmed to be independent from the location, though the small but measurable decrease in the curvature was observed near the edge of highly curved samples. Approximately 1.5 mm length and 100 m width surface profile was recorded along with the centerline, in which more than three scatter-free lines were averaged for determining the annealed and/or irradiated curvature of each specimen.

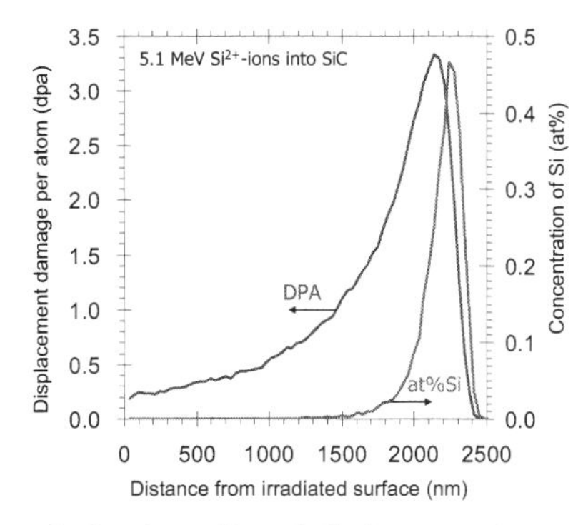

Figure 2. Depth profiles of displacement damage and concentration of Si for the case of nominal damage of 1 dpa in -SiC calculated by SRIM 98.

RESULTS AND DISCUSSION

Surface strain determination

The surface curvatures of SiC samples irradiated without stress were plotted in Fig. 3 as a function of linear swelling reported previously [2], in which lacking or some scattered swelling data in [2] were re-examined in this work. The irradiated curvatures of each sample thickness are proportional to the linear swelling as shown in Fig. 3. The proportional constants of 0.02, 0.009, and 0.005 were obtained for the samples with thickness of 50, 75, and 100 microns, respectively. The irradiated curvature for non-stressed cases is likely attributed to the strain mismatch between unirradiated and irradiated regions. Under the condition of the limited thermal creep rates, the strain was introduced only in the irradiated region for the following possible reasons; (1) the irradiation induced swelling, (2) the irradiation creep associated with the compressive stress due to the swelling, and (3) mean lattice dilation associated with the Si-ion retention. If the irradiation creep rate induced by the compressive stress is proportional to both the (swelling rate) and $(\text{stress})^n$ (the exponent of n•1) as was traditionally found in irradiated metals, the final creep rate should be roughly proportional to the $(\text{swelling rate})^{(n+1)}$, because the compressive stress can be assumed as nearly proportional to the swelling. However, the irradiated curvature in figure 3 showed the nearly linear relationship with the linear swelling. Although, the strain rate due to the ion retention is proportional to the irradiation time or ion-flux, the swelling is not the case. Therefore, the plastic strains associated with (2) and (3) were likely small comparing to the strain by swelling in the conditions studied here. This is probably due to the stress mitigation at the boundary between unirradiated and irradiated regions. The both free ends of the irradiated surface, those are not constraint toward a longitudinal sample direction, may allow the some mechanisms for the stress mitigation such as localized deformation and reorganization of the atomic bonding.

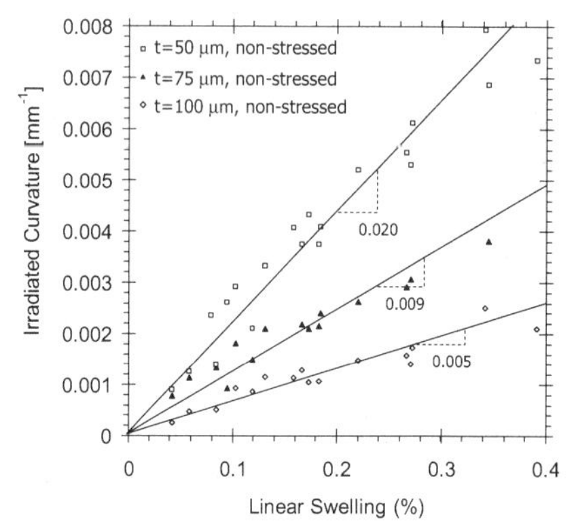

Figure 3. The relationship between irradiated curvature and linear swelling for the cases of irradiation without stress.

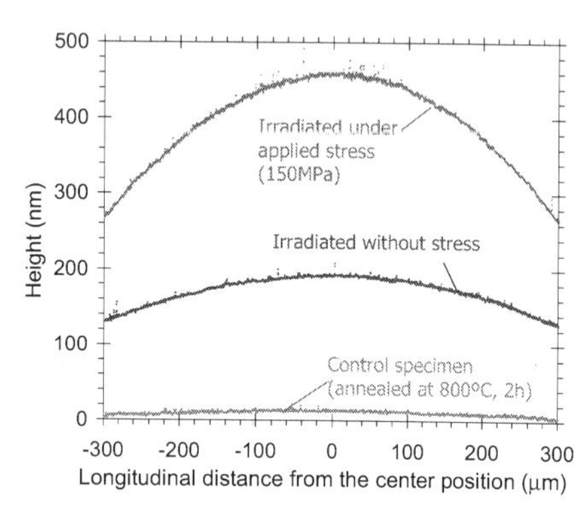

Figure 4. Results of the profilometry of the sample surface using atomic force microscopy following ion-irradiation at 800°C, 3dpa with or without stress. The results for the control specimen annealed at 800°C for 2h was also shown.

AFM profiles of the irradiated or unirradiated (annealed) surfaces of samples with 50 m thickness were shown in Fig. 4, where green line corresponds to the edge view of control specimen annealed at 800°C for 2 h, blue line shows the sample irradiated at 800°C to 1 dpa without applied stress, and red line shows the sample irradiated at 800°C to 1 dpa with applied stress of 150 MPa, respectively. The incremental curvature for the sample irradiated with applied stress (red line) comparing to the irradiated curvature of unloaded sample (blue line) likely associates with the stress related additional strain in the stress axis. Hereafter, the creep strain is defined as the differential strain between the longitudinal sample strain for the stress irradiation cases and that of non-stress irradiation cases. In the case such that the irradiated curvature is proportional to the longitudinal strain as demonstrated for the non-stressed cases, the additional surface strain or the irradiation creep strain may be written using the obtained proportional constant of A by

$$\varepsilon = \frac{1}{A}\left(\kappa_{AS} - \kappa_{th}\right) \qquad (1)$$

, where κ_{AS} is the curvature of the sample irradiated with applied stress and κ_{th} is the curvature of control specimen with the same thickness as irradiated samples.

Relationship with swelling

Figure 5 shows the relationship between the irradiation creep strain obtained from the eq. (1) and the non-stressed linear swelling. The irradiation creep strain increased with increasing in the linear swelling, where the linear dependence indicates that the irradiation creep rates are proportional to the swelling rates. The proportional constants were strongly dependent on the irradiation temperature, for example the highest constant of 1.43 for 800°C was approximately one order of higher than that of 0.15 observed for 400°C. In the temperature regime between 400-800°C, the swelling was reported to be logarithmically proportional to the DPA below ~1 dpa and it reached the salutation values [2]. Note that the irradiation creep strain still follow the swelling even in the saturation swelling regime as shown in Fig. 5. The creep rates between 1 and 3 dpa were roughly estimated to be 10^{-8}-10^{-7} [MPa^{-1}-dpa^{-1}] above 1 dpa. The swelling saturation values were reported to decrease with increasing in the irradiation temperature, the lowest value was found at ~800°C. By contrast, the highest rate of the "transient" irradiation creep was observed at 800°C. The damage levels at which the swelling becomes saturate showed good accordance with the damage levels at which the population of microstructural defects practically saturates, and the temperature dependency on the swelling saturation levels was best explained by the accumulation of microstructural defects, especially point defects and its tiny clusters, in this temperature regime [1, 8, 9]. Thus, the linear relationship clearly shown in Fig. 5 and the temperature dependence of the proportional constants implicate that the irradiation creep in SiC is also related to the microstructural defects. Above the temperature studied here or at least above 1000°C, void swelling may commence due to somewhat long range migration of the vacancies [10]. Because the reduced sink density should strongly accelerate the mobility of mobile defects such as self-interstitials, the dominating creep mechanisms would not be same as below 800°C and the linear creep-swelling relationship may not be expected at above 1000°C. In addition to this, the contribution of thermal creep is also expected at higher temperatures.

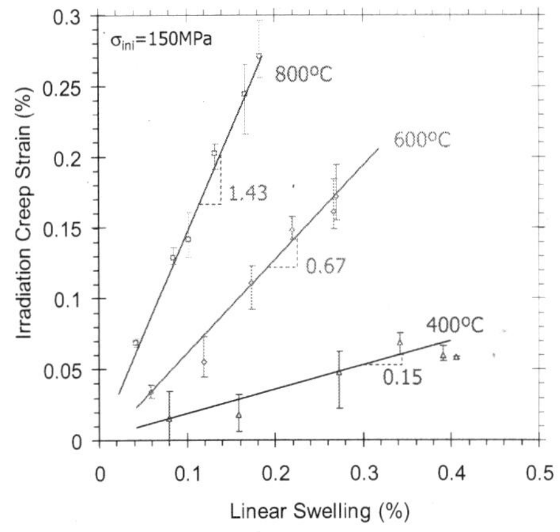

Figure 5. The relationship between irradiation creep strain of the -SiC irradiated under applied stress (150 MPa) and linear swelling. The linear swelling [2] was measured from the samples irradiated with Si-ions without stress.

Stress dependence

Figure 6 shows the relationship between the irradiation creep and stress for the irradiation temperatures of (a) 400°C, (b) 600°C, and (c) 800°C, respectively. In all the temperatures, the irradiation creep strain showed nearly linear dependence on the stress at all the damage levels ranging 0.01-3 dpa. The creep strain rate is traditionally denoted as proportional to the stress ^{n} as stated above, where the stress exponents of irradiation creep generally ranging from 1 to 5 were reported for metals except for the cases with substantial stress levels [11]. Although the creep rates seem to have been strongly dependent on the stress at the early stage of the irradiation, the dependence was practically absent above 0.01 dpa for SiC.

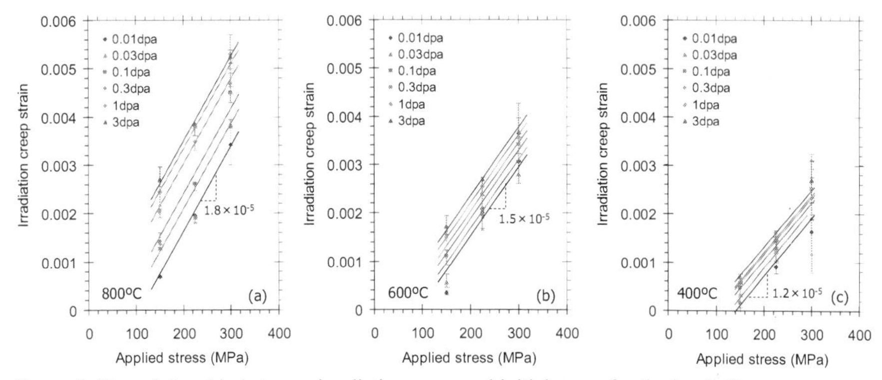

Figure 6. The relationship between irradiation creep and initial stress for the irradiation temperature of (a) 800°C, (b) 600°C, and (c) 400°C. The proportional constants indicated in each charts were overall average of plots for all damage levels.

SUMMARY

High purity CVD SiC samples were irradiated with Si^{2+}-ions to 0.01-3 dpa at 400, 600, and 800°C under applied stress. The stress was applied by means of bending the thin samples during irradiation and the initial applied stresses at the irradiated surface were 150, 225, and 300 MPa, respectively, depending on the sample thickness. Creep strains were estimated from the irradiated sample curvatures based on comparisons with the curvatures of strips irradiated under non-stressed conditions. Irradiation creep strain was strongly dependent on the displacement damage and irradiation temperature, where the approximate linear relationship with swelling measured for non-stressed SiC was found for all the irradiation temperatures. The creep strain also showed nearly linear relationship with the applied stress level. The creep equation which is similar to the traditional irradiation creep equation developed for metals was obtained.

REFERENCES

[1]L.L. Snead, T. Nozawa, Y. Katoh, T.S. Byun, S. Kondo, and D.A. Petti, Handbook of SiC properties for fuel performance modeling, *J. Nucl. Mater.*, **371**, 329-377 (2007).
[2]Y. Katoh, H. Kishimoto, and A. Kohyama, The influences of irradiation temperature and helium production on the dimensional stability of silicon carbide, *J. Nucl. Mater.*, **307-311**, 1221-1226 (2002).
[3]L.L. Snead, Y. Katoh, and S. Connery, Swelling of SiC at intermediate and high irradiation temperatures, *J. Nucl. Mater.*, **367-370**, 677-684 (2007).
[4]N.B. Morley, Y. Katoh, S. Malang, B.A. Pint, A.R. Raffray, S. Sharafat, S. Smolentsev, and G.E. Youngblood, Recent research and development for the dual-coolant blanket concept in the US, *Fusion Eng. and Des.*, **83**, 920-927 (2008).

[5]Y. Katoh, L.L. Snead, T. Hinoki, S. Kondo, and A. Kohyama, Irradiation creep of high purity CVD silicon carbide as estimated by the bend stress relaxation method, *J. Nucl. Mater.*, **367-370**, 758-763 (2007).
[6]Proceedings of the 3rd International Congress on Ceramics, Osaka, Japan (2010).
[7]A. Kohyama, Y. Katoh, M. Ando, and K. Jimbo, A new Multiple Beams-Material Interaction Research Facility for radiation damage studies in fusion materials, *Fusion Eng. and Des.*, **51-52**, 789-795, (2000).
[8]Y. Katoh, N. Hashimoto, S. Kondo, L.L. Snead, and A. Kohyama, "Microstructural Development in Cubic Silicon Carbide during Irradiation at Elevated Temperatures," *J. Nucl. Mater.,* **351**, 228-240 (2006).
[9]S. Kondo, Y. Katoh, and L.L. Snead, Microstructural Defects in SiC Neutron Irradiated at Very High Temperatures, *J. Nucl. Mater.*, **382**, 160-169 (2008).
[10]R.J. Price, Neutron irradiation-induced voids in -silicon carbide, *J. Nucl. Mater.*, **48**, 47-57 (1973).
[11]M.L. Grossbeck, and J.A. Horak, Irradiation creep in type 316 stainless steel and us PCA with fusion reactor He/dpa levels, *J. Nucl. Mater.*, **155-157**, 1001-1005, (1988).

Joining and Integration of Ceramic Structures

PRELIMINAR RESULTS ON JOINING OF THIN SiC/SiC COMPOSITES BY SILICIDES COMPOUNDS AND LOCAL HEATING

Elodie Jacques[1], Laurence Maillé[1], Yann LePetitcorps[1], Christophe Lorrette[1, 2], Cédric Sauder[2]

[1]Université Bordeaux 1, Laboratoire des Composites ThermoStructuraux, UMR 5801, 3 allée de la Boétie, Pessac, France
[2]CEA Saclay DEN/DMN/SRMA/LTMEX, Gif sur Yvette, France

ABSTRACT

The present work describes the methodology and the results for joining SiC substrates in the solid state by reactive and non reactive metallic silicides in order to use these materials as container for the Gas Fast Reactor. The operation temperatures are in the range of 1000°C but short time incursions at 1600 or 2000°C have to be anticipated. As joining material, we have chosen to study four titanium and niobium silicides: $TiSi_2$, $NbSi_2$ (non reactive with SiC), Ti_5Si_3 and Nb_5Si_3 (reactive with SiC) in order to modify the joint/substrate bonding. The joining between these silicides and the silicon carbide substrate was prepared by Spark Plasma Sintering in order to limit the contact time during the sintering. Post thermal treatments in the solid state performed under a controled atmosphere have been carried out in order to study the thermal stability of the assembly. Polished cross sections were prepared to study the interface composition, the cracking path in the vicinity of the joint/substrate interface. In order to decrease the residual stresses within the joint, SiC particles were mixed with the silicides powders. In the case of the non reactive silicides, the SiC particles effectively reduce the cracks whereas for the reactive silicides, the SiC particles react with the joint leading to a thermally stable composition and an improved bonding to the substrate. In order to prepare the joining technology, few trials of local heating have been investigated with a CO_2 laser beam. The advantage of local heating is the possibility to heat at high temperature (around 1800°C) for short time without damaging the composite structure. Thermal measurements have been done from one side to the other of the silicon carbide plate.

INTRODUCTION

For forty years, the energy production never stopped growing to respond the demographic and industrial growth. Today, controversies concerning the use of fossil energies and the poor yield of renewable energies have made nuclear power essential. It represents 80% of the energy production in France. The ageing of nuclear power plants encourages countries all around the world to devote a part of their research programs to the Generation IV reactors. Such reactors have to meet technology goals: improve nuclear safety, minimize waste and natural resource utilization, and to decrease the cost for building and running such plants. The French institution CEA (Commissariat à l'Energie Atomique) focuses its efforts on different concepts of reactors and in particular on the Gas Fast cooled Reactor (GFR). This reactor is helium-cooled, with an outlet temperature in the range of about 500°C to 1000°C in normal situation. Several fuel forms are being considered for their potential to operate at very high temperatures and to ensure an excellent retention of fission products, examples given for ceramic matrix composite (CMC) tubes or honeycombs, advanced fuel particles, or ceramic clad elements of actinide compounds. In this study, we focused on a tube concept made of thin CMC: SiC/SiC composites. [1]

Silicon carbide ceramic matrix composites have been developed at first for applications in severe environments such as rockets, jet engines or gas turbines. But the temperature requirements for Generation IV reactors make them attractive also for nuclear applications. In actual reactors, metallic alloy tubes are used as fuel container but such alloys have too poor properties above 500°C. Silicon carbide (SiC) has sufficient thermal and mechanical properties: a low coefficient of thermal expansion ($4.10^{-6}K^{-1}$), a good thermal conductivity ($30W.m^{-}$

$^{1}.K^{-1}$ at room temperature for CVD silicon carbide), a high stiffness (400GPa) and high decomposition temperature (until 2500°C in inert atmosphere).[2] Moreover, SiC is neutron transparent in order to allow good performances of the nuclear core. The structural materials are also submitted to different thermomechanical strains: thermal gradients, helium pressured at 70 bars in the reactor... and then need to have good damage tolerance, thermal shock resistance and a limited brittleness. Mechanical properties of unreinforced silicon carbide are noticeable with a Young modulus twice as much as steel at room temperature but its brittleness is important. Using SiC fiber reinforcement in a SiC matrix permits to increase the damage tolerance (typically 0.6% of tensile deformation before break at room temperature) thanks to a damage mechanism. SiC/SiC ceramic composites are then good candidates to answer all these specifications.

Nevertheless, there are some technological issues to overcome such as the difficulty to accomplish complex geometries. Then, a suitable method of joining SiC_f/SiC_m components is required. Riveting and bolting are not considered for ceramics and the key issue resides in joining pieces by brazing, soldering (not possible on very refractive materials such as silicon carbide) or solid state diffusion. The joining material should have the same thermomechanical properties as the composite and have to keep these properties at very high temperature. Several other joining methods have been developed for the assembly of ceramics such as mechanical fastening, plasma spray or physical and chemical vapor deposition. Brazing is used extensively due to its simplicity, low cost investment and potential to be a mass production process. In recent years, joining SiC with Ag-Cu or Cu-based braze has been reported but their utilization is restricted to 500°C. Other authors have developed eutectic brazes from metals but the presence of free silicon doesn't enable their use beyond 1400°C. Many studies have been carried out on Ti-Si-C and Cr-Si-C systems on eutectic compositions: Si-22%wtTi ($TiSi_2$ and Si), Si-18%atCr ($CrSi_2$ and Si) and Si-44%atCr (CrSi and $CrSi_2$). [3][4][5]

The aim of this study is to find relevant compositions and processing routes for joining SiC/SiC thin composites. The requirements for the joining composition can be describe as follow : the joint has to be adherent the the silicon carbide substrate with a limited reaction zone in order to avoid the damaging of the SiC, the joint must not be cracked in order to avoid gas leaks, the assembly must withstand for short durations temperatures around 1600°C. Taking into account these requirements, four silicides, available in a powder shape, have been selected ($TiSi_2$, Ti_5Si_3, $NbSi_2$, Nb_5Si_3). They have a higher CTE and a lower Young's modulus than SiC. In order to reduce the strain mismatch, few amount of SiC powders have been added to the joining composition. Moreover, M_5Si_3 (M=Ti or Nb) silicides are reactive to SiC, and in order to limit the reaction with the silicon carbide substrate, few amount of SiC powders or metallic dilisicide powders (MSi_2) have also been added to the joining composition. The joining of the two silicon carbide substrates was prepared in the solid state by Spark Plasma Sintering in order to limit the contact time during the sintering. Post thermal treatments in the solid state performed under a controled atmosphere have been carried out to study the thermal stability of the assembly. Polished cross sections were prepared to observe the interface composition and the cracking path in the vicinity of the joint/substrate interface.

MATERIALS AND METHODS

<u>Selection of the constituents</u>

The study of ternary diagrams (M, Si, C) shows a chemical stability of some metal-based silicides or carbides with silicon carbide, whereas other silicides of the same metal are unstable and forms a metallic carbide (Figure 1.a-c). By considering the matches of the properties between SiC and the joining material, MSi_2, M_5Si_3 and MC (M=Ti or Nb) can be considered as potential joining compounds (Table1).

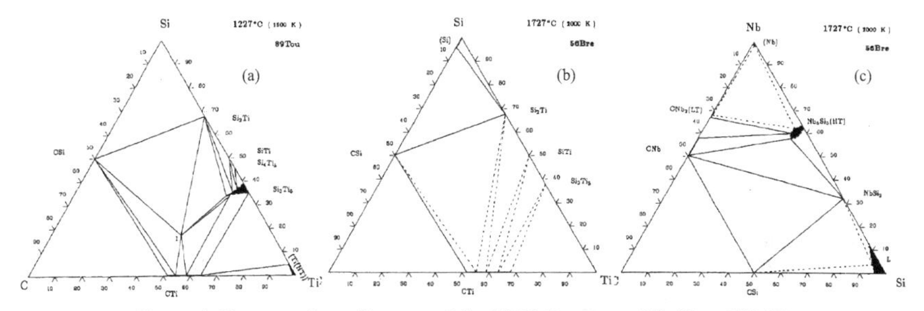

Figure 1. Ternary phase diagram of the M-Si-C systems (M=Ti or Nb) [6]
(a) Ti-Si-C at 1227°C, (b) Ti-Si-C at 1727°C, (c) Nb-Si-C at 1727°C

Table 1. Physical properties of compounds of interest [7] [8]

	SiC	TiC	NbC	Ti_3SiC_2 (T_1)	$TiSi_2$	Ti_5Si_3	$NbSi_2$	Nb_5Si_3
Tmp (°C)	2700	3140	3480	3000	1540	2120	1930	2480
α (10^{-6} K^{-1})	4,5	7,6	6,8	9,3	10,4	11	8,4	7,3
E (GPa)	400	450	340	325	260			

The ternary phase diagrams of Ti-Si-C and Nb-Si-C show the thermodynamic stability of disilicides with the silicon carbide. On the contrary, with silicides such as Ti_5Si_3 or Nb_5Si_3, a reaction zone should be observed due to the absence of equilibrium. Along the diffusion paths of the metallic elements, the following reaction zones may be produced (Figure 1):

Ti_5Si_3-> TiC -> SiC (1)
Ti_5Si_3-> $TiSi_2$ -> SiC (2)
Ti_5Si_3 -> TiC -> T_1 -> SiC (3)
Nb_5Si_3 -> NbC -> $NbSi_2$ -> SiC (4)
Nb_5Si_3 -> NbC -> SiC (5)

The main advantage of using chemically stable silicides as joining material is the chemical stability in time of the assembly. But with this lack of chemical reaction, a weak bonding between the elements can be expected. On the other hand, a strong bonding is attended with the M_5Si_3 silicides but to the prejudice of a chemical damage of the SiC substrate. A compromise has to be found among adhesion, mechanical behavior against cracks and chemical reactivity with silicon carbide.

By considering the thermomechanical behavior, the mistmatch between the physical properties of SiC and the silicides shows that after cooling, the SiC is in a compression state, whereas the silicides are in tension. Due to the lack of room temperature ductility of the silicides, the joint will be cracked. In order to decrease the CTE and to increase the rigidity of the joint, 0 to 20%vol. of SiC were added to all the silicides. In the case of the M_5Si_3 silicides, the chemical reaction with SiC will transform the joint composition into a more stable one as described before. The different compositions of the joint are summarized in Table 2.

Table 2. Compositions of interest for the joint (M=Ti or Nb)

	Reasons of choice	Expected behavior after sintering	Expected behavior after a post-thermal treatment
MSi_2	Chemical stability	Cracking within the joint	Better sintering
MSi_2 + % $SiC_{powders}$ 0-20%vol.	Decrease the residual stresses	Less cracking according to the volume fraction of reaction zone	Stable
M_5Si_3	Increase of the bonding with the SiC	Better bonding and reaction zone with SiC	Stable composition but with a damaging of SiC
M_5Si_3 + % $SiC_{powders}$ Vo. : 0-20%	Limit the reaction zone with the SiC	Better bonding and lower reaction zone with SiC	Stable composition with a controlled reaction of SiC
MSi_2+M_5Si_3+$SiC_{powders}$ (40-40-20)%vol.	Limit the reaction zone with the SiC and decrease in the residual stresses	Better bonding, lower reaction zone with SiC, no cracks in the joint	Stable composition with a controlled reaction of SiC

Preparation of the samples

Powders of 99.5% $TiSi_2$ (-325 mesh, Alfa Aesar), 99.5% Ti_5Si_3 (-325 mesh, Alfa Aesar), 99.85% $NbSi_2$ (-325 mesh, Alfa Aesar), 99.5% Nb_5Si_3 (-325 mesh, Alfa Aesar) and pure silicon carbide (around 1μm, CEA) were used as starting materials. The purity of the powders was controlled by XRD analysis. The different ratios of the powders were prepared according to the table 2. The mixture of the powders was performed for 8h using an automatic stirrer.

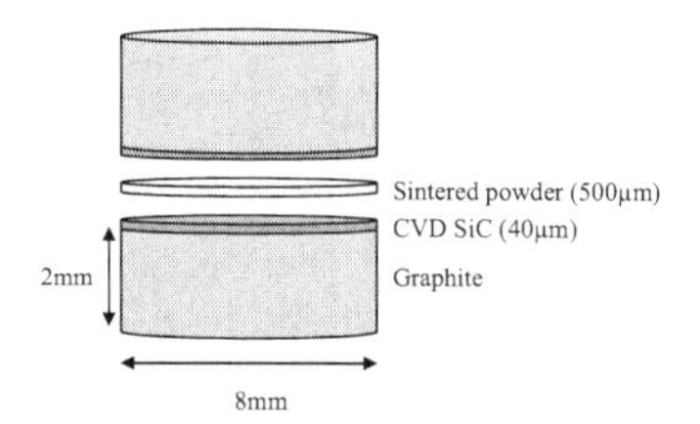

Figure 2. Drawing of the prepared samples

After mixing, the powders were placed in a graphite die (inside diameter, 8mm) between to pellets of graphite (reference 2175 from Mersen) covered by CVD SiC (thickness 40μm) and introduced into a Spark Plasma Sintering equipment (Figure 2). This device permits the sintering of ceramics in a very short time. Limiting thus the grain growth and the reaction zone with the substrate. The singularity of this technique is the application of a uniaxial pressure on the sample during the process combined with an alternative or DC current. The pressure in the chamber is around 5Pa. The sintering temperature is always 200°C below the melting temperature of the compound in order to maintain a solid state reaction. Thus, Table 3 summarizes the sintering temperatures of the prepared samples. The temperature rising rate was 100°C/min and a pressure of 50MPa was applied since the sintering temperature is reached. A 5 min step is applied at the sintering temperature and at the end of the process, the samples were cooled to room temperature with a cooling rate of 100°C/min. The morphology, the microstructure and the chemical composition of the as prepared samples were examined by scanning electron microscopy (SEM), Jeol with energy dispersive X-ray analysis (EDAX) and X-ray diffraction (XRD), Brucker.

Table 3. Sintering temperatures of the prepared samples

	Temperature (°C)
$TiSi_2$ + % $SiC_{powders}$ Vo. : 0-20%	1300
$NbSi_2$ + % $SiC_{powders}$ Vo. : 0-20%	1600
Ti_5Si_3 + % $SiC_{powders}$ Vo. : 0-20%	1900
Nb_5Si_3 + % $SiC_{powders}$ Vo. : 0-20%	1900
$TiSi_2$+Ti_5Si_3+$SiC_{powders}$ (40-40-20)%vol.	1300
$NbSi_2$+Nb_5Si_3+$SiC_{powders}$ (40-40-20)%vol.	1600

Thermal treatments at 1600°C for 3h were carried out on all the samples except for $TiSi_2$ + % $SiC_{powders}$ Vo. : 0-20% because of a melting point temperature below 1600°C. These treatments were performed in a furnace heated by an inductor in a controled atmosphere of argon. The heating and cooling rate were 20°C/min. The treated samples were also examined by SEM and XRD.

RESULTS AND DISCUSSION

Non reactive silicides (MSi_2 or MSi_2+SiC)

We have first studied the sintering of the chemically stable silicides $TiSi_2$ and $NbSi_2$. The good adhesion of the two is relevant but cracks are observed in both cases (Figure 3a). Those cracks are very harmful while they are crossing grains and even within the SiC substrate. To minimize this effect, we introduced a certain amount of SiC fine powders. For $TiSi_2$, the SEM images confirm the former hypothesis while the higher the amount of SiC fine powders is the smaller density of cracks is observed (Figures 3a-c). No thermal treatment has been carried out on $TiSi_2$ as its melting point temperature is below 1600°C.

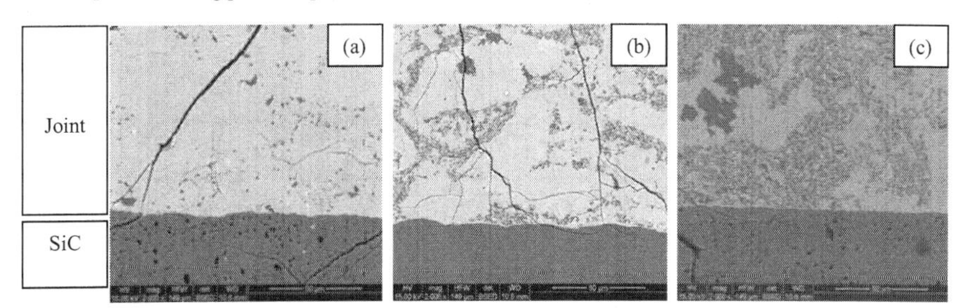

Figure 3. SEM micrographes of as-pressed materials (the SiC particles are the grey spots) (a) $TiSi_2$, (b) $TiSi_2$ + $10\%_{(vol)}$ SiC, (c) $TiSi_2$ + $20\%_{(vol)}$ SiC

A good adhesion is observed for $NbSi_2$+20%SiC (Figure 4a) with cracks within the joint. Even if the global morphology of the joint is acceptable, a worsening is noted after 3hours at 1600°C (Figure 4b) with a small decohesion in some places.

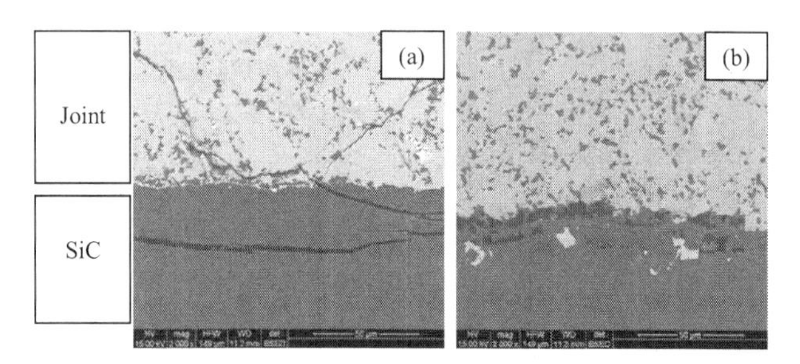

Figure 4. SEM micrographes of as-pressed materials (the SiC particles are the grey spots) (a) $NbSi_2$+20%SiC as pressed, (b) $NbSi_2$+20%SiC after 3h at 1600°C

To conclude on these compositions, although MSi_2 + 20%$SiC_{particles}$ are not enough refractive to keep a good morphology at 1600°C for 3h, these silicides are interesting materials on a physico-chemical point of view. Indeed this thermal treatment is very constraining and represents an extreme case. In order to improve the joint morphology, it would be better to use smaller silicides powders (similar to SiC ones), this would permits also to reduce the joint thickness. The assembly was done under pressure, it would be interesting to study the bonding without any pressure during the sintering to conclude on the cracks.

Reactive silicides (M_5Si_3 or M_5Si_3+SiC)

The same experiments have been carried out with the reactive silicides Ti_5Si_3 and Nb_5Si_3 (Figure 5, 6 and 7).

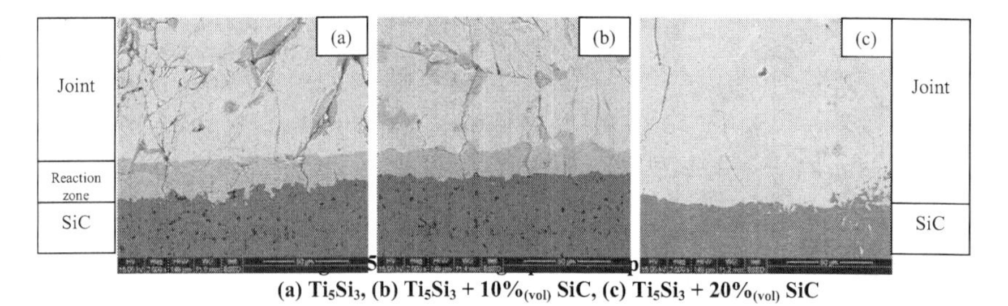

(a) Ti_5Si_3, (b) Ti_5Si_3 + 10%$_{(vol)}$ SiC, (c) Ti_5Si_3 + 20%$_{(vol)}$ SiC

The presence of a reaction zone between the joint and the SiC substrate is observed when the joint contains less than 10% of SiC_p (Figure 5. a,b). As expected, there is a chemical reaction at the interface with the silicon carbide substrate. For higher content of SiC_p (20 or 30%), the reaction occurs at the interface but also within the joint itself (Figure 5c). The chemical transformation of the joint seems to be correlated to the amount of SiC powder incorporated. With 0 to 10% of SiC, less than 5% of the joint thickness has reacted, 20% are transformed with 20% of SiC_p and nearly 35% are transformed with 30% of SiC_p. The EDS analysis of this reacted zone reveals the presence of titanium carbide (TiC_x) against the SiC substrate whereas no $TiSi_2$ is observed. Then the diffusion path is represented by the equation (1). The presence of TiC_x is interesting for at least two reasons, TiC has a Young's modulus and a CTE close to SiC and TiC_x is in equilibrium with SiC and Ti_5Si_3 (Figure 1, Table 1).

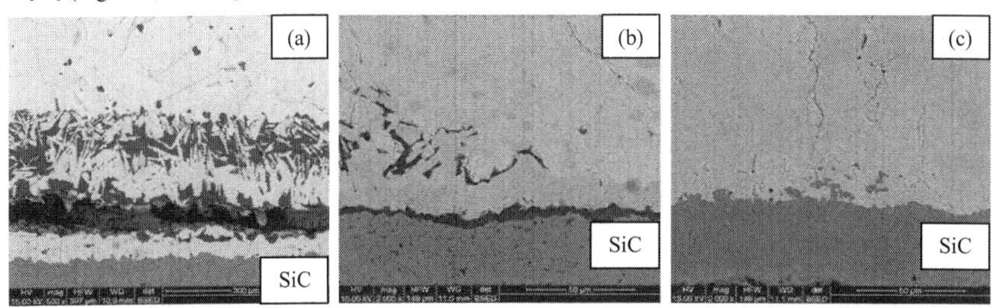

Figure 6. SEM micrographes after 3h at 1600°C heat treatment
(a) Ti_5Si_3 , (b)Ti_5Si_3 + 10%$_{(vol)}$ SiC, (c) Ti_5Si_3 + 20%$_{(vol)}$ SiC

After thermal treatment, a large debonding and a strong reaction with the SiC substrate is noticeable (Figure 6 a-b) for the low content of SiC_p (0-10%vol). These results emphasize the role played by the SiC powders mixed to the silicide (Figure 6c). An EDS analysis has revealed that the thermal treatment leads to the formation of the Ti_3SiC_2 phase within a Ti_5Si_3 matrix and is coherent with the ternary phase diagram (Figure 1a). The diffusion path after thermal treatment corresponds to the equation (3). The presence of the pseudo-ductile Ti_3SiC_2 is interesting for mechanical purpose of the assembly. After 3h hours of diffusion time, the joint is still composed 65% of unreacted Ti_5Si_3 in the middle of the joint thickness. The Ti_3SiC_2 formed at the joint/ silicon carbide interface plays the role of a diffusion barrier.

In the case of Nb_5Si_3 with a large amount of SiC_p, a good behavior is also observed before and after a thermal treatment (Figure 7a-b). Within the joint, the thickness of the reaction zone is limited by comparison with Ti_5Si_3 whereas the substrate is strongly chemically attacked (Figure 7a-b). The EDS analysis leads to the formation of niobium carbide (NbC_x). The chemical composition of Nb_5Si_3 after thermal treatment remains the same with the presence of the NbC_x interface between SiC and Nb_5Si_3. Whereas NbC_x and TiC_x are very similar with a vacancy in carbon, the difference of chemical reaction between the two silicides with SiC can be explain by the presence of the T_1 compound (in the Si,C, Ti) phase diagram which can play the role of a diffusion barrier. As a consequence even if Nb_5Si_3 has a higher melting point than Ti_5Si_3, it is less interesting for this application because of a more important chemical composition modification in time than the Ti-Si-C system. After the thermal treatment at 1600°C, some un-reacted Ti_5Si_3 is still present in the joint and the composition of the blend ($SiC+Ti_5Si_3$) has to be tuned.

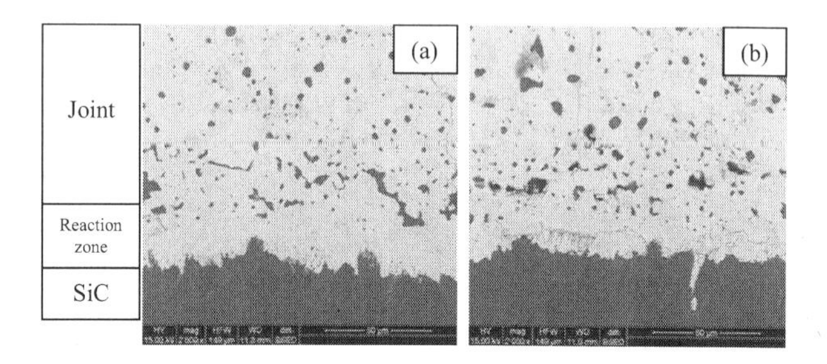

Figure 7. SEM micrographes of Nb_5Si_3+20%SiC joint composition (a) as-pressed, (b) after 3h at 1600°C heat treatment

Non reactive and reactive silicides (MSi_2 + M_5Si_3+SiC (40/40/20%vol.))

In order to get rid of the drawbacks of reactive and non reactive silicides and above all to minimize the presence of the reaction zone, we have blended them in a certain quantity and observed the behavior of the joints (Figure 8 and 9).

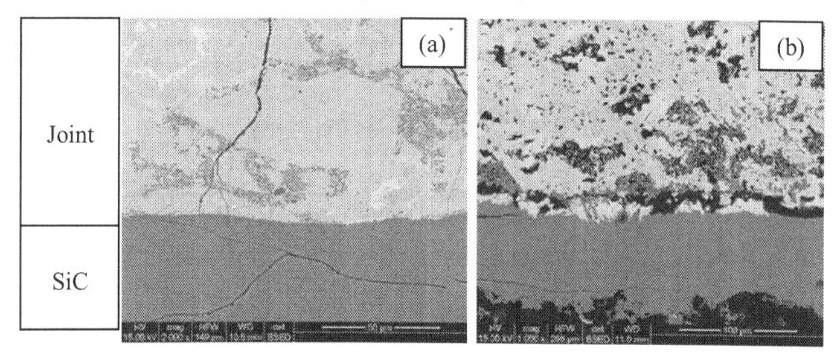

Figure 8. SEM micrographes of 40$TiSi_2$ + 40Ti_5Si_3 + 20SiC(% Vo.)/ SiC interfaces (a) as-pressed , (b) after 3h at 1600°C heat treatment

For $TiSi_2$+Ti_5Si_3+SiC composition, a lack of reaction zone is relevant even after a severe heat treatment. Few cracks are however present and are crossing the joint from the SiC substrate showing a good bonding. The ternary phase diagram of the Ti-Si-C system shows that $TiSi_2$ and Ti_5Si_3 are not stable and should form TiSi (Figure 1). We don't observe the formation of this phase but the presence of TiC at the interface of Ti_5Si_3 grains. Then, we assume that Ti_5Si_3 reacts with the SiC fine powder more easily than with the disilicides to form TiC_x and probably Ti_3SiC_2. Due to the presence of this stable phase $TiSi_2$, the quality of the joint after a thermal treatment is strongly degradated because of the melting of $TiSi_2$ at 1600°C. But as said above, the treatment performed is very constraining and the requirements for this application concerns few minutes at 1600°C instead of 3h. Nevertheless, the chemical behavior of such joint is interesting and the starting powder sizes of $TiSi_2$ and Ti_5Si_3 have to be decreased in order to have a thinner joint. Moreover, by decreasing the size of the powders, it will make an easier sintering.

For $NbSi_2$+Nb_5Si_3+SiC composition, we observe the formation of a reaction zone between the joint and the substrate but it is considerably thinner than in the case Nb_5Si_3 alone (25µm against 5µm in that case (Figures 7a and 9a)). However, the thermal treatment leads to very large and damageable cracks in the joint and at the interface between the substrate and the joint where the NbC_x is present (Figure 8c).

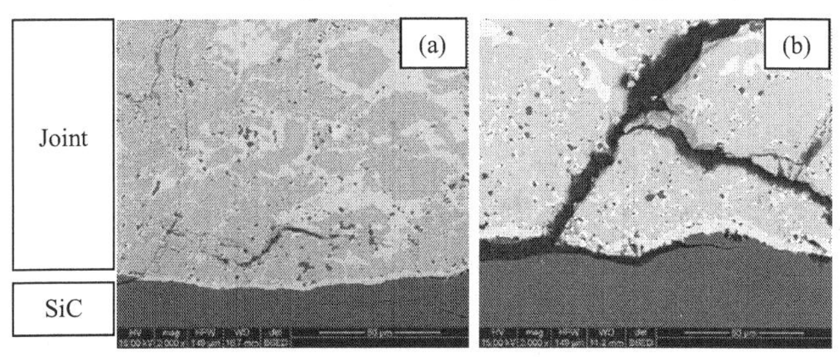

Figure 9. SEM micrographes of 40$NbSi_2$ + 40Nb_5Si_3 + 20SiC(% Vo.)/ SiC interfaces (a) as-pressed , (b) after 3h at 1600°C heat treatment

Local heating with a CO_2 laser beam

The former experiments were performed with SPS equipment. Such process is interesting to obtain sintered joints at high temperature with a good compaction for short time. It limits the atom diffusion between the joint and the SiC substrate and then a chemical modification of the assembly during the sintering. But a pressureless process is required for assembling thin SiC/SiC composites. Indeed, these materials have a better tensile deformation than ceramics but a uniaxial pressure at high temperature on thin composites would be very damageable. Moreover, the thermomechanical properties of SiC fibers are strongly reduced for thermal treatment over 1800°C but a rapid stay at this temperature can lead to limited damages. The joining process has to be rapid, local and pressureless. The laser feats these requirements by offering a heating zone delimited by the laser beam and an operating temperature obtained quasi instantaneously.

Heating experiments were performed using a CO_2 laser (λ=10,7µm and f=1000Hz). The Figure 10. describes the device with an horizontal oval laser beam (4x15mm diameter). Plates of SiC/SiC composites (thickness 1mm) were vertically fixed on a sample holder in order to be in front of the laser beam. The sample holder was placed in a vacuum vessel in order to control the atmosphere during the heating and avoid a strong oxidation of the composite. The temperature of the front and the back faces of the samples were measured by two pyrometers (one in the range 600-1700°C and the other 800-2500°C). Two types of thermal profiles were performed : one with one plate to evaluate the thermal comportment of the SiC/SiC composite under the laser beam (1) and the other by putting together two plates to simulate a joining process (2).

The Figure 11. describes the thermal profiles in the two configurations. The laser power was modified by step of 15W every minute from 60W to 180W, which leads to a heating rate of 100°C/min for a single plate (case (1)). The thermal results (Figure 11 (1)) outlines a linear increase of the temperature on the two faces of the plate as a function of the laser power applied. Besides a difference of 200°C between front face and back face of the plate is continuously observed from 1000 to 1800°C. In the case of two plates putted together (Figure 11 (2)) a different behavior is observed. Indeed, the thermal profile of the face receiving directly the laser beam is no longer linear. At low temperature (typically 1000°C), the heating is faster than at high temperature (typically

1800°C). The other face hardly warm up and when the front face is at 1800°C, a 1000°C gap is observed with the other face. It can explain the thermal profile of the front face which is 'cooled' by the other face due to an important thermal gradient between the two (thickness 2mm and limited contact between the two plates).

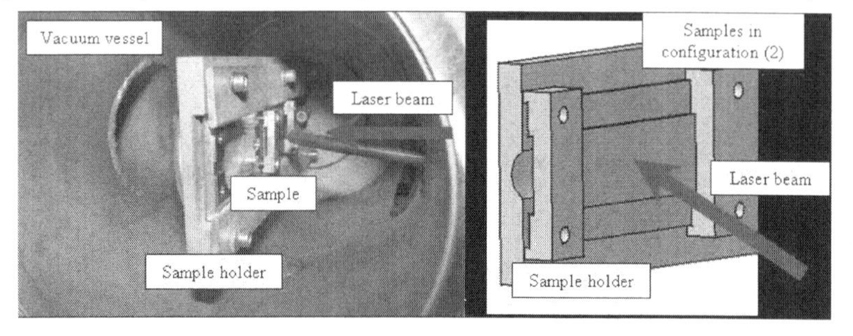

Figure 10. Schemas of the device and the sample holder

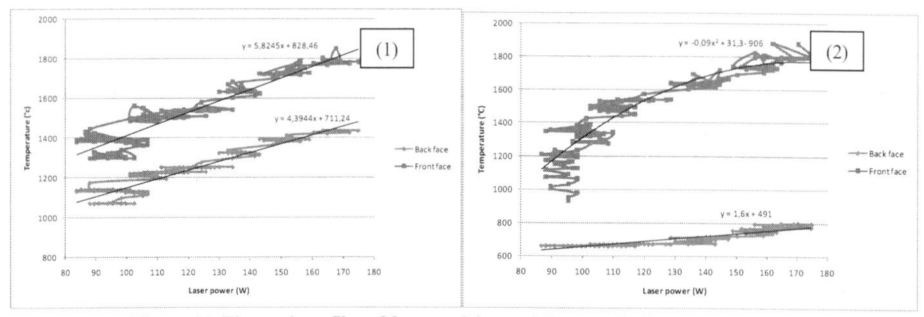

Figure 11. Thermal profiles of front and faces of the samples in cases (1) and (2)

To conclude on these experiments, it seems that the case (2) is not valide for assembling composites. The back face is too isolated from the front face by the thickness of the assembly and the limited contact of the two composites, roughs in essence. The joint has to face directly the laser beam to be heated enough otherwise it will be impossible to reach the attended temperatures for the joining process. By modifying the sample holder, this heating process is a promising one to join thin SiC/SiC composites.

CONCLUSION

Whatever the composition of the silicide is, it is necessary to blend the silicide powder with fine SiC particles (1 m). The amount of SiC has not yet been adjusted but at least 20% in volume are necessary. A compromise has to be find : in order to minimize the residual stresses a maximum of SiC is required but with a high amount of SiC_p, the sintering and the solid state diffusion to the substrate will be more difficult. For all the composition, in order to facilitate the sintering and to decrease the thickness of the joint, the initial powder sizes of the silicides have to be decreased. Three types of composition were found to be of interest : (1) The first one is a

$TiSi_2$+SiC, the stability is limited to the melting temperature of the disilicide (1540°C) but this composition can be sintered at a low temperature and present a good adhesion, (2) the second one Ti_5Si_3+SiC is unstable, TiC_x and Ti_3SiC_2 are formed at the substrate interface allowing the bonding without damaging the substrate, the amount of the Ti_5Si_3 has to be tuned in order to have a stable joint in time, (3) Finally, the 3rd composition $TiSi_2$+Ti_5Si_3+SiC is of interest because the presence of $TiSi_2$ limits the quantity of Ti_5Si_3 reacting with the silicon carbide substrate. The operating temperature is however limited to the melting temperature of the disilicide. The adhesion of these composition has now to be tested by mechanical tests. Local heating tests have been carried out using a CO_2 laser on thin SiC/SiC composites and give promising results for further joining experiments.

ACKNOWLEDGEMENTS

Financial support: MATINEX
Laser experiments: B. Guizard, H. Haskrot, CEA Saclay, Gif sur Yvette
SPS experiments: C Estournès, G. Chevallier, PN²F, Toulouse

REFERENCES

[1]: Innovative SiCf/SiC composite Materials for Fast Reactor Application, L. Chaffron *et al.*, Matériaux 2010, Nantes

[2] : Rôle des interfaces dans le brasage non réactif du SiC par les siliciures de Co et de Cu, A. Gasse, Thèse Université de Grenoble, France (1996)

[3] : Low activation brazing materials and techniques for SiCf/SiC composites, B. Riccardi, C.A. Nannetti, T. Petrisor, M. Sacchetti, Journal of Nuclear Materials 307–311 (2002) 1237–1241

[4] : High temperatures brazing for SiC and SiCf/SiC ceramic matrix composites, B. Riccardi, C.A. Nannetti, T. Petrisor, J. Woltersdorf, 2002, International Symposium on SiCf/SiC composites materials research and development

[5] : Issues of low activation brazing of SiCf/SiC composites by using alloys without free silicon, B. Riccardi, C.A. Nannetti, T. Petrisor, J. Woltersdorf, Journal of Nuclear Materials 329–333 (2004) 562–566

[6] : SGTE diagrams

[7] : Materials Handbook, A concise Desktop Reference, F. Cardarelli, Springer, ISBN 1-85233-043-0

[8] : Shear fracture behavior of Ti_3SiC_2 induced by compression at temperatures below 1000°C, Z.F. Zhang, Z.M. Sun, Materials Science and Engineering 408 (2005) 64–71

JOINING OF NITE SiC/SiC COMPOSITE AND TUNGSTEN FOR NUCLEAR APPLICATIONS

H. Kishimoto[1], T. Abe[1], T. Shibayama[2], J.S. Park[3], H.C. Jung[3], Y. Kohno[1], A. Kohyama[1]

[1]Muroran Institute of Technology, Muroran, Hokkaido 050-8585, Japan
[2]Hokkaido University, Sapporo, Hokkaido 060-8628, Japan
[3]Muroran Establishment, IEST Co., Ltd., Muroran 050-8585, Japan

ABSTRACT

Silicon carbide (SiC) is an expectant candidate of the structural material for the next generation nuclear systems. For the production of a component, many parts of SiC/SiC composites and metals will be assembled by mechanical and metallurgical joining techniques. The joining technique of dissimilar materials with SiC/SiC composite is an important issue to establish a nuclear system. Nano-Infiltration and Transient Eutectoid (NITE) method is a recently developed fabrication technique which is able to produce a large, complex shape parts of SiC/SiC composite. There are many metals as opponents for the joining of dissimilar materials with the SiC/SiC composite. For the first wall of blanket and the armor on the divertor in fusion reactor, tungsten (W) is a candidate metal for the joining. Because tungsten with NITE SiC/SiC composites will be used at high temperature near or over 1000 °C, the diffusion bonding is appropriate for this joining. A large advantage of the W and SiC joining is their very closed coefficients of thermal expansion (CTE). In present research, seeking and developments of the joining techniques of dissimilar materials focusing on tungsten for nuclear system are researched.

INTRODUCTION

Silicon carbide (SiC) is an expectant candidate of the structural material for the next generation nuclear systems such as Gas-cooled Fast Reactor (GFR) systems and fusion reactors after Demonstration Power Plant (DEMO). Nano-Infiltration and Transient Eutectoid (NITE) method is a recent progressed processing method to fabricate SiC/SiC composite. The NITE method is able to provide dense, large, complex shaped parts of SiC/SiC composite[1]. The parts of SiC will be assembled into a component, then, joining techniques of dissimilar materials are essential. Utilization of the SiC/SiC composites mainly aims to increase the energy efficiency by the increase of the operation temperature, thus, a joining technique for the elevated temperature is the most essential. Tungsten (W) is one of the most frequently utilized opponent materials with SiC in the designs of nuclear system. In the case of the fusion reactor, SiC is a candidate for the structural material of the first wall and the divertor. Tungsten armor is planning to be employed to protect the SiC construction of the divertor. For the fission reactors, SiC is a candidate for cladding tubes in the helium gas-cooled fast reactor, and tungsten is possible to be used for the upper-end plug of the fuel pin. Joining methods such as brazing, reaction sintering, using polymer or zinc-borate glass have been researched[2-5]. There are many techniques to produce SiC/SiC composites such as PIP, CVI NITE and reaction sintering methods. The NITE process fabricates composites under high pressure and high temperature over 1700 °C resulting

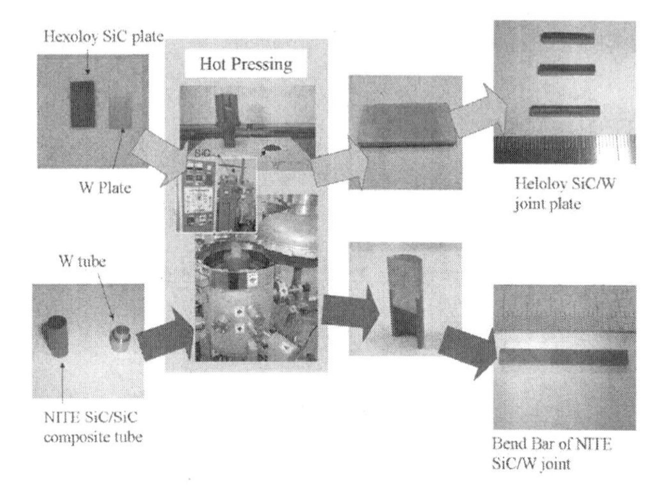

Figure 1. Outlines of fabrication process of SiC/W joints

in having dense matrix and stable microstructure at elevated temperature, thus, the NITE SiC/SiC composite is able to resist against the high pressure and temperature of the diffusion bonding. The diffusion bonding is appropriate for high temperature systems which will employ the SiC/SiC composite as structural material. The coefficient of thermal expansion (CTE) of tungsten (4.6 x 10^{-6}/ K) is very close to that of beta-SiC (4.7 x 10^{-6}/ K), the joints of SiC and tungsten do not need insert materials to absorb the mismatch of CTE. Present research produces NITE-SiC/SiC composites and monolithic SiC plate joints with tungsten, and investigates the interface and fracture surface using mainly microstructural investigation methods.

EXPERIMENTAL

SiC materials used were both monolithic SiC plates and NITE SiC/SiC composites. The monolithic SiC plates were Hexoloy™ SiC provided by Hitachi Chemical Co., Ltd.. The Hexoloy SiC is an alpha-SiC produced by the pressure-less sintering. The NITE-SiC/SiC composite was provided by IEST Co., Ltd.. The matrix in NITE SiC/SiC composite was beta-SiC produced by the hot isostatic pressing (HIP) from SiC nano-powders. The used tungsten was cold-worked polycrystalline plate of 99.9% purity. The contact surfaces of the plates were polished with diamond powders. The diffusion bonding was performed by hot-pressing at 1600°C at 20 MPa in an argon atmosphere using a hot press device (Hi-Multi 5000, Fujidempa Co. Ltd). Holding times at the hot-pressing temperature were investigated from 0.5h to 10h. Figure 1 shows the outline of the processing. The SiC/W joints were cut and sliced to be bending specimens with the dimension of 26 x 3 x 1 mm. The mechanical strength of the joint was measured by 3-point bending with support span of 18 mm at room temperature. The crosshead speed of the testing was 0.5 mm/min. Microstructures were examined using a field-emission

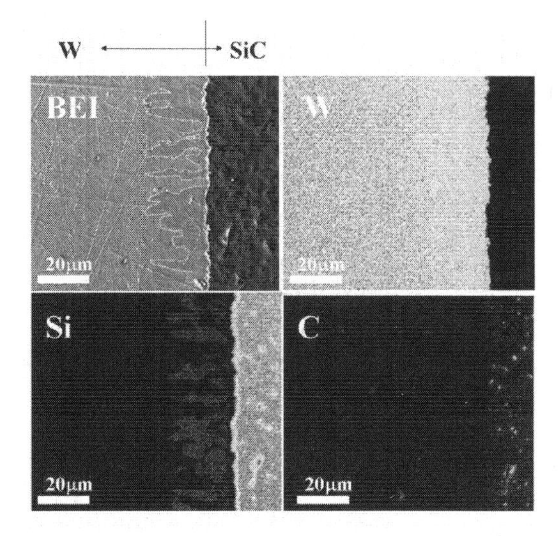

Figure 2. Back scattering electron image (BEI) and EPMA maps of interface between Hexoloy SiC and tungsten plates fabricated at 1600°C, 20MPa for 10h

scanning electron microscope (FE-SEM, JEOL JSM6700F), an electron probe micro analyzer (EPMA, JEOL JXA-8900R).

RESULTS

Figure 2 shows a back scattering electron images and EPMA maps for a joint of monolithic Hexoloy SiC and tungsten plate fabricated at 1600°C and 10h. On the back scattering electron image in Figure 2, column like dark contrasts extend from the interface into tungsten. The dark contrasts are indicated by white-dot lines on the back scattering image in Figure 2. The EPMA analysis reveals that the contrasts are the diffusion of silicon in SiC into tungsten. For the investigation of the diffusion behavior of NITE SiC/SiC composites and tungsten, the EPMA analysis was performed. A SEM image and EPMA maps are shown in Figure 3. A bundle of SiC fibers are selected for this analysis, and any silicon diffusions into tungsten were not observed in this area. Apparent reaction layers could not be detected by this EPMA analysis[6]. To investigate the reason of the difference, the interface between the NITE SiC/SiC composite and tungsten plate was observed by the back scattering electron images as shown in Figures 4(a), (b) and (c). Figure 4 (a) is the interface between SiC matrix and tungsten, the column like dark contrasts surrounded with black dot lines were observed. Figure 4 (b) is a boarder of SiC matrix and a SiC fiber bundle, and the dark contrast seems to be disappeared. 3-point bend testing for the joints of NITE SiC/SiC composite and tungsten was performed and the back scattering electron images of the fracture surface are shown in Figure 5. The flexural strength of the joints was between 100MPa and 160MPa. Flexural strain of 0.05-0.1% was observed. The SiC side fracture surface in

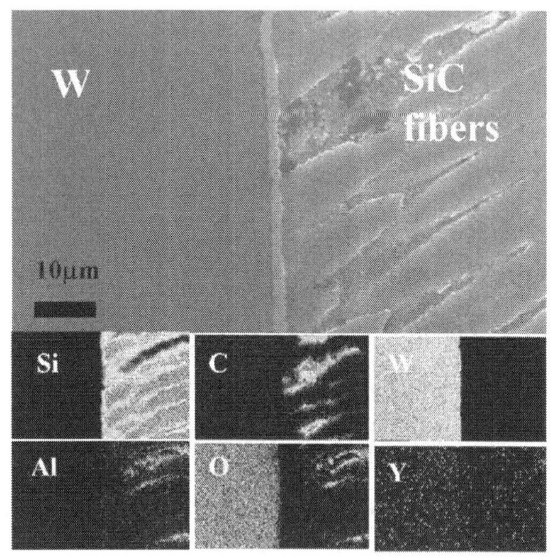

Figure 3. SEM image and EPMA maps of interface between NITE SiC/SiC composite and tungsten fabricated at 1600°C, 20MPa for 1h

Figure 5 has uniform contrast, however, the tungsten side fracture surface is divided into white and black contrast areas. The lower Z elements are shown darker contrast in the back scattering electron images, the white area is tungsten and dark area is estimated to be SiC. Both SiC matrix and fibers are put on the tungsten fracture surface, thus, the crack is considered to propagate interface and within SiC.

DISCUSSION

Diffusion bonding between SiC plate and tungsten are normally driven by the diffusion of silicon and carbon. In the case of the joint of monolithic Hexoloy SiC and tungsten plates, silicon diffusion into tungsten and formation of tungsten silicide having column like shapes were observed. The bonding strength of this joint is considered to be produced by the diffusion. The Hexoloy SiC is simple, but the NITE SiC/SiC composite has complex construction and chemical contents. The composite consists of beta-SiC matrix and SiC fibers with carbon coating, some residual sintering additives of alumina and yttria. These oxides sometimes affect the bonding procedure[7]. The EPMA analysis at a fiber bundle in Figure 3 could not detect any apparent diffusion of tungsten which is observed in Figure 2. There is possible that very thin oxide layer of several tens nano-meter thick formed between the SiC fiber and tungsten, but it is still not certain. Figure 4 shows that the diffusion of silicon into tungsten occurred in the NITE SiC/SiC composite as same as the joint of Hexoloy SiC with tungsten, Main mechanism of NITE SiC/SiC composite with tungsten joint is the silicon diffusion into tungsten but more precise microstructural analysis using TEM is necessary to analyze the interface construction,

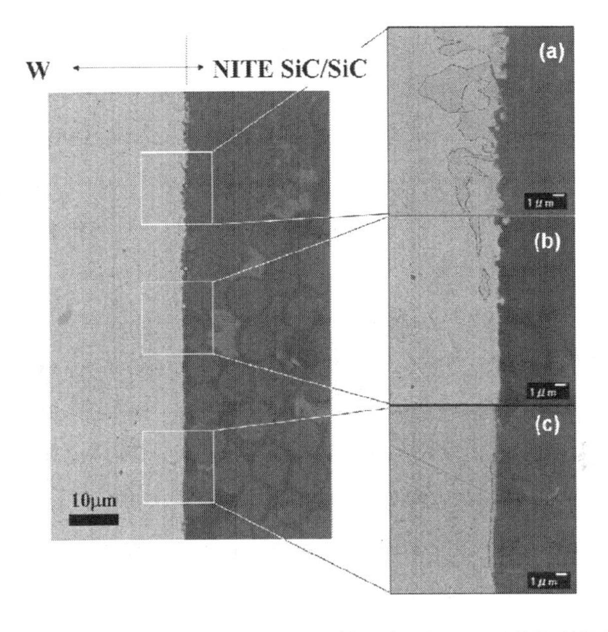

Figure 4. Back scattering electron images of interface between NITE SiC/SiC composite and tungsten fabricated at 1600°C, 20MPa for 1h.

especially, in the SiC bundle area. The fracture surface indicated that the crack propagated in the NITE SiC/SiC composite resulting in SiC putting on the tungsten fracture surface. Both matrix and fibers were left on the tungsten, each SiC fibers seemed to be joined to tungsten in spite of no apparent diffusion behavior. The flexural strength of 3-point bend testing was over 100MPa, the joining seemed to be successfully established.

CONCLUSION

Research of joining for the NITE-SiC/SiC and tungsten by diffusion bonding aiming to a nuclear application was tried using the hot-pressing method. Microstructural investigation using SEM and EPMA showed that both Hexoloy SiC plate and NITE SiC/SiC composite were joined with tungsten plate mainly by the diffusion of silicon into tungsten. The silicon diffusion behavior was not appear near SiC fiber bundle despite that the each SiC fibers were well joined to tungsten, The other joining mechanisms, such as the formation of thin oxide layers, is considered to work to join the SiC fibers to tungsten. Flexural strength of the joints reached to over 100MPa, the diffusion bonding for the NITE-SiC/SiC and tungsten was successfully established in present research.

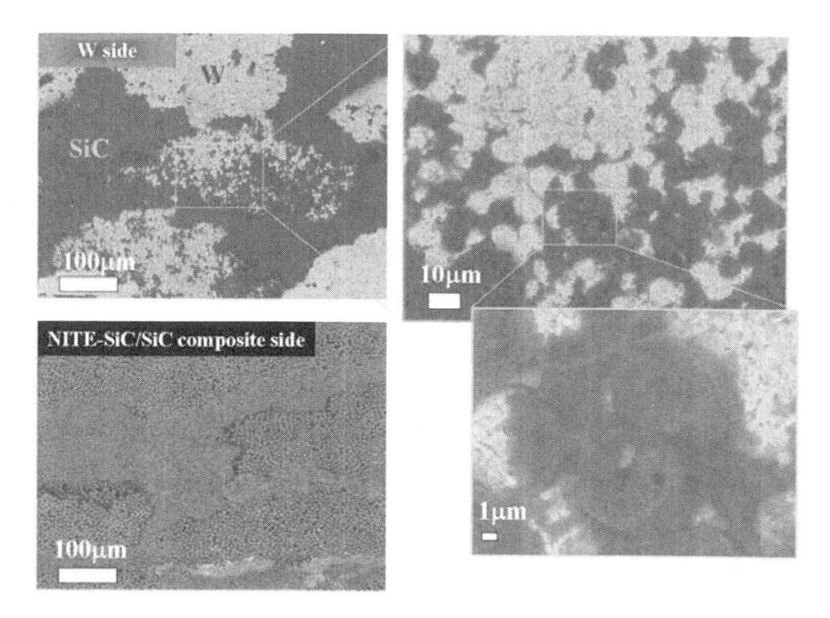

Figure 5. Back scattering electron images of fracture of NITE SiC/SiC composite and tungsten after 3 point bend testing

ACKNOWLEDGEMENTS

The present study is partially the results of "Development of nano-mechanics joining analysis technology in advanced materials for gas cooling fast reactor, Design and development of a nano-mechanics joining analysis device" entrusted to Hokkaido University by the Ministry of Education, Culture, Sports, Science and Technology of Japan (MEXT), and collaboration research in the project of "Advanced Materials R & D by Means of Complex and Ultimate Environmental Effect Evaluation Methodology", so called FEEMA project, entrusted to Muroran Institute of Technology by the Ministry of Education, Culture, Sports, Science and Technology of Japan (MEXT).

REFERENCES

[1] A. Kohyama, T. Hinoki T. Mizuno, J.S. Park, H. Kishimoto, Advanced GFR utilizing NITE-SiC/SiC Shield Fuel Pin, *Proceedings of the 2006 International Congress on Advances in Nuclear Power Plants, ICAPP'06*, 2149-2156 (2006).

[2] B. Riccardi, C.S. Nannetti, T. Petrisor, J. Woltersdorf, E. Pippel, S. Libra, L. Pilloni, Issues of low activation brazing of SiC_f/SiC composites by using alloys without free silicon, *J. Nucl. Mater.*,

329-333, 562-566 (2004).
[3] C.H. Heneger, Y. Shin, Y. Blum, L.A. Giannuzi, B.W. Kempshell, S.M. Schwarz, Coatings and joining for SiC and SiC-composites for nuclear energy systems, *J. Nucl. Mater.*, **367-370**, 1139-1143 (2007).
[4] P. Lemoine, M. Ferraris, M. Salvo, M.A. Montorsi, Vitreous Joining Process of SiCf/SiC Composites, *Journal of the European Ceramic Society*, **16**, 1231-1236 (1996).
[5] P. Colombo, B. Riccardi, A. Donato, G. Scarinci, Joining of SiC/SiC_f ceramic matrix composites for fusion reactor blanket applications, *J. Nucl. Mater.*, **278**, 127-135 (2007)
[6] H. Kishimoto, T. Shibayama, T. Abe, K. Shimoda, S. Kawamura, A. Kohyama, Diffusion Bonding Technology of Tungsten and SiC/SiC Composites for Nuclear Applications, *Proceedings of 3rd International Congress on Ceramics (ICC3), Nov. 14-18 2010, Osaka Japan*, submitted.
[7] H. Kishimoto, T. Shibayama, K. Shimoda, T. Kobayashi, A. Kohyama, Microstructural and Mechanical Characterization of W/SiC Bonding for Structural Material in Fusion, *J. Nucl. Mater.*, in press.

Processing

INTEGRATED R & D OF SIC MATRIX CERAMIC COMPOSITES FOR ENERGY/ENVIRONMENTAL APPLICATION

Akira Kohyama[1,2], Y. Kohno[1], H. Kishimoto[1], J. S. Park[1,2], H. C. Jung[1,2] and K. Shimoda[1]

1:OASIS, Muroran Institute of Technology
2: MuroranEstablishment, IEST Co., Ltd.
27-1 Mizumoto-cho, Muroran 050-8585, JAPAN

kohyama@mmm.muroran-it.ac.jp

ABSTRACT

Integrated R & D efforts for industrialization of advanced SiC/SiC composites and SiC based composites by fabrication system integration of NITE method are on-going under Organization of Advanced Sustainability Initiative for Energy System/Material (OASIS), at Muroran Institute of Technology. As one of the activities at OASIS, a prototype production line of green sheets and prepreg sheets has been recently installed. The near-net shaped preforms with the NITE green-sheets and prepreg-sheets are made into near-net shape components for potential applications under consideration by HIP and Pseudo-HIP. Aiming at the near term utilization, SiC/SiC hybrid structures with metallic materials, such as steels and other refractory metals, were fabricated with promising results. Trial to reduce the production cost has been continued. Also, many options for bonding and joining SiC/SiC with metallic materials and FRP have been investigated. Those results are provided.

1. INTRODUCTION

Muroran Institute of Technology, well known national institute of technology with legendary materials R & D activities, is locating in city of Muroran, south-west of Hokkaido Island. On March, 2009, Facility of Energy and Environmental Material Assessment (FEEMA) was established to initiate and promote SiC/SiC related activity in Muroran Institute of Technology. Followed to the strong start up activities of FEEMA, Organization of Advanced Sustainability Initiative for Energy System/Materials (OASIS) was established March, 2010. As shown in Fig.1, OASIS consists of 4 divisions and FEEMA facility [1]. As is indicated, OASIS emphasizes its efforts on energy system/Material R & D where SiC/SiC and SiC based innovative materials R & D is the core and major activity as the leading efforts in Japan.

SiC fiber R & D, started with the PCS-type SiC fiber development at IMR, Tohoku University early 70th has been continued with the shift from metal matrix composite R & D for super-sonic transport (SST) to hyper-sonic transport (HST), then SiC fiber reinforced composites with metallic, C and SiC matrix composites. Those were mainly supported by METI (ministry of economy, trade and industries) and the authors' group has been involved changing the place from University of Tokyo, Kyoto University to Muroran Institute of Technology. The continuing efforts for almost two decades

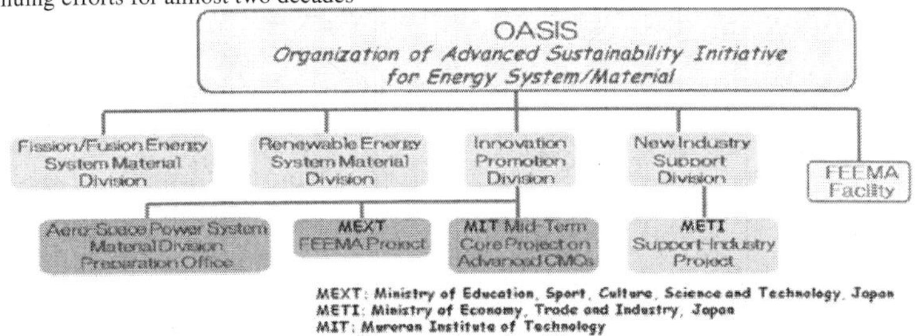

Fig. 1: Structure of OASIS at Muroran Institute of Technology

on SiC/SiC R & D, which includes invention of NITE (nano-infiltration and transient eutectic phase) method made many progresses in nuclear materials and aero-space materials engineering [2].

The strong efforts for industrialization of NITE method have been carried out under the collaboration of Institute of Energy Science and Technology (IEST) Co., Ltd., and Kohyama Laboratory of IAE, Kyoto University. The major part of Kohyama Laboratory of Kyoto University was moved to Muroran Institute of Technology March 2009. Now this collaboration is expanding in Muroran between OASIS, Muroran Institute of Technology and Muroran Technical Establishment of IEST Co., Ltd.,

2. PROJECT STREAM AND JESSICA

SiC related activities under OASIS include Mid-term core program on advanced CMC funded by Muroran Institute of Technology, support industry program funded by METI and some tasks of FEEMA Project, as shown in Fig. 1. The SiC/SiC R & D in Muroran Institute of Technology is based on the authors' patents, NITE method (as shown in Fig. 2). The SiC related project of the MIT's mid-term core program is called as **S**iC **T**echnology **R**enaissance for **E**nergy and **A**erospace in **M**uroran (STREAM), which has a main object to establish prototype production line of NITE-SiC/SiC mid-products, such as green sheets and prepreg-sheets. By using those mid-products, SiC/SiC preforms are produced. Those are hot pressed, hot isostatically pressed or pseudo-isostatically hot pressed to produce near net shape products or simple plate or block products, as shown in Fig. 2.

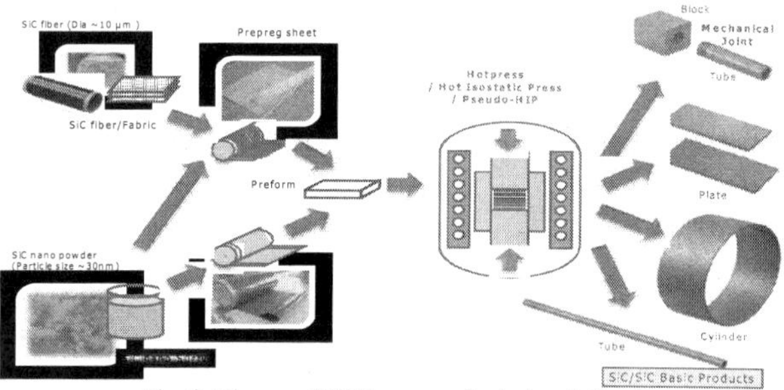

Fig. 2: The new NITE process for industrialization

The properties are tailored by material design of fiber types, fiber architecture, fiber-matrix inter-phase types, inter-phase geometry and matrix types. The project of the support industry program of METI is "Development of Die-cast machine by hot-chamber method with utilizing advanced SiC/SiC as core component". The final goal of this project in three years is to complete and put into the market the new Die-cast machine with sufficient economical competitiveness and technological attractiveness.

In order to promote those activities, joint venture activity called JESSICA (**J**oint Venture for **E**nergy and

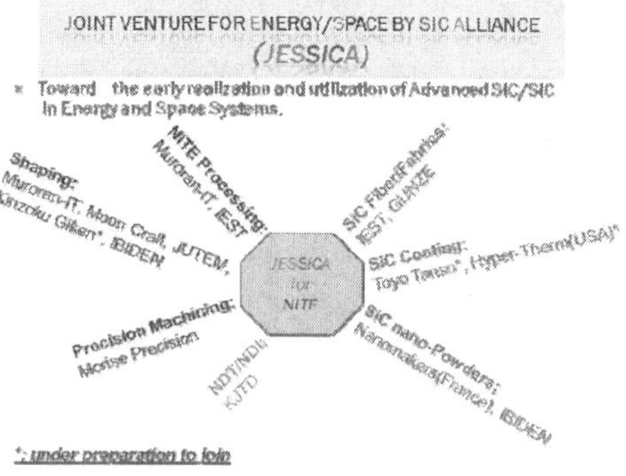

Fig. 3: Structure of JESSICA

Aero-Space by **SiC** Alliance) has been initiated including industries and universities which cover from mother materials to final products and evaluation/inspection. The current structure of JESSICA is shown in Fig. 3. This is the so-called integrated R & D of SiC Matrix Ceramic Composites for Energy/Environmental Application

3. NEW NITE PROCESS FOR INDUSTRIALIZATION

The nano-powder infiltration and transient eutectic phase (NITE) is an advanced liquid phase sintering method, where slurry with SiC nano-powders and quite small amount of sintering agents is infiltrated into SiC fibers to become SiC matrix by high temperature sintering process. In the original process invented, the slurry was produced by using water. The water based slurry has a difficulty to keep preform shape while handling this into making final product. For industrialization, large scale production capability with easy, reliable and stable process is essentially required. As the solution, dry and flexible mid-products are designed. In the new process, as shown in Fig. 4, slurry is using polymers as solvent and produce green-sheets and prepreg-sheets as mid-products.

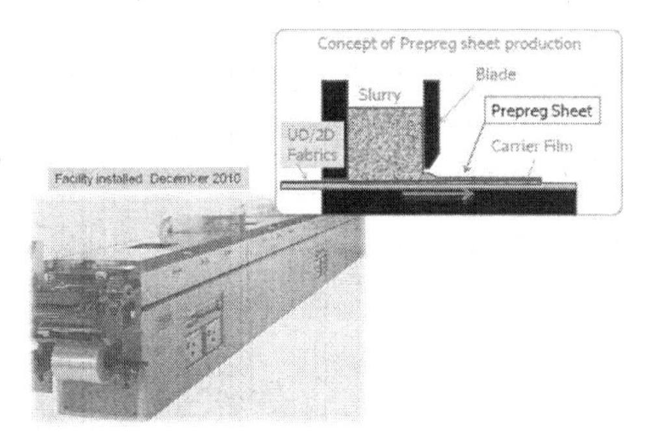

Fig. 4: NITE-SiC/SiC prepreg-sheets prototype production line at OASIS

Prepreg-sheet is SiC fibers with SiC slurry infiltrated and green sheet is SiC slurry in flexible sheet shaped. A prototype production machine of the NITE-SiC green-sheets and prepreg-sheets was installed December, 2010 in OASIS, Muroran Institute of Technology, as shown in Fig. 4. The green-sheets is foamed on 40cm width carrier film with the maximum speed of 40cm/min continuously. The prepreg-sheets are also produced with a slight modification of starting section. The process establishment of preform formation to final products is still under preparation. However, the current industrialization plan does not strictly require the preform production process shift from full hand-made to semi-automatic process. This level of process improvement and integration will be and should be done by industries, neither by OASIS nor universities.

4. NITE-SIC/SIC PRODUCTS FROM OASIS

In the past, tubes, plates and even scale models of fusion blanket, IHX, nuclear fuel, fuel pins and other nuclear system components were fabricated by the original NITE-method. In these five years the efforts to move to new NITE process with green-sheet and prepreg-sheets were done and now new products are being made from OASIS under partly collaboration with IEST Co., Ltd. One direction is to make multi-layered SiC/SiC with monolithic SiC layers at the surface, which provide excellent environmental resistance, ware resistance and many other new features. NITE SiC/SiC composites are tailored materials and their fiber architecture and inter-phase/matrix/fiber structure are designed based on their material requirements. Figure 5 is the typical example to change matrix structure in the new NITE process. In this case, matrix structure is full β-SiC or (β+α)-SiC. Currently, most of the products are full β-SiC structure with pyro-C inter-phase.

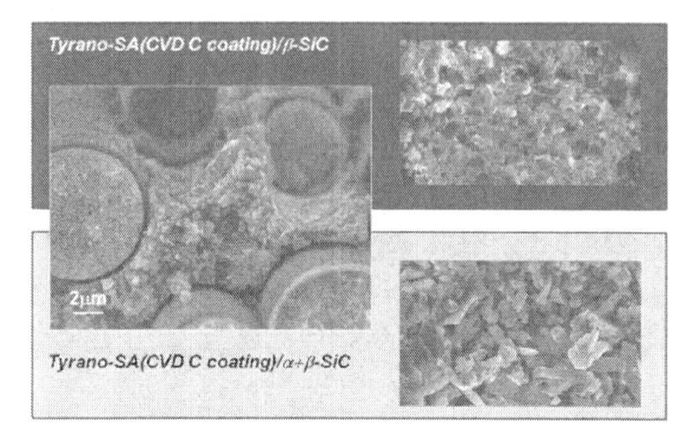

Fig. 5. Representing structures of NITE-SiC/SiC with different matrix structure

Although SiC/SiC composites are very attractive as highly environmental resistant materials, the high material cost is a serious problem. In order to reduce cost burden, utilization of inexpensive amorphous fibers has been planned in many ways. In-service crystallization of amorphous SiC causes detrimental effects on system performance and high crystallinity of fibers and matrix of SiC/SiC is absolutely necessary for the fission and fusion applications. Thus, in-situ crystallization of amorphous SiC fiber during SiC/SiC production by NITE-method has been developed, as IC-NITE process. Radiation damage behavior under heavy ion irradiation of SiC/SiC by IC-NITE process was provided with presenting its potential as nuclear fusion/fission materials [3]. IC-NITE SiC/SiC was made with NITE green-sheets and NITE prepreg sheets utilizing amorphous SiC fibers coated with phenolic resin.

In Fig. 6, cross sectional view is indicating the well controlled fiber architecture with pyrolized carbon as intra-fiber-bundle matrix. Fracture surface after bending test includes sufficient fiber pull-outs indicating good pseudo-ductility of the material. Also, well crystallized fiber structure with fell densified matrix can be seen [4].

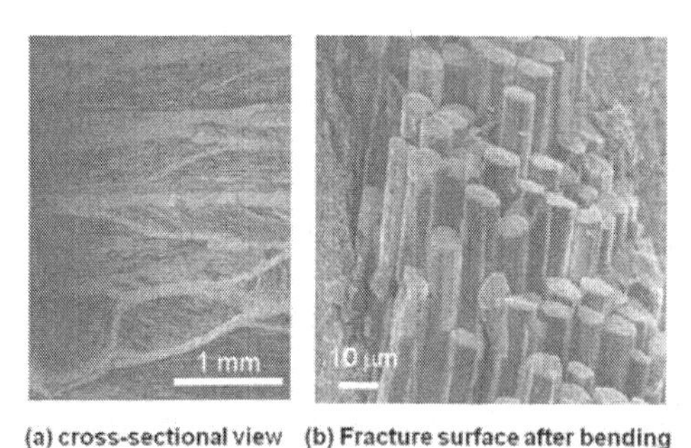

Fig. 6: IC-NITE SiC/SiC for low cost, high performance application

One of the important applications of IC-NITE is the on-going METI 's support industry program under collaboration with GUNDAI Co., Ltd., where IC-NITE SiC/SiC is applied for die cast machine with hot chamber method. The current die cast machine with its technical issues and the R & D items are shown in Fig. 7. In this R & D activity, multilayered SiC/SiC and cast ion and SiC/SiC hybrid materials are also designed and are fabricated. The main technical challenges are (1) Accurate machining with smooth surface, (2) Hardness control for mechanical seal and (3) Large Cast-ion/SiC/SiC Component fabrication by shrink fitting. The advanced design of the die-cast machine should present sufficient lifetime of cylinders, leak tightness of mechanical seal, function of floating bulbs and heating capability of hot chamber by induction heating.

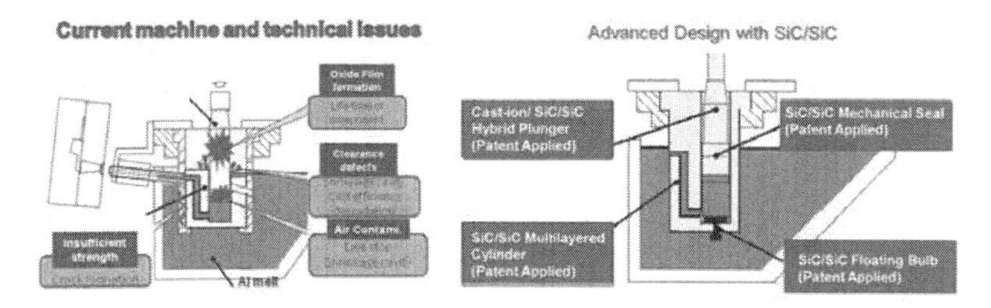

Fig. 7. On-going R & D of new NITE technology application to die-cast machine

SiC/SiC multilayered cylinders and SiC/SiC mechanical seals have been successfully fabricated by in-situ crystallization process. Figure 8 indicates outlooks of the preform and final products through pseudo HIP process and final shaping, where well controlled fiber architecture can bee seen. The sizes of the cylinder cover 8mm to 50 mm in diameter.

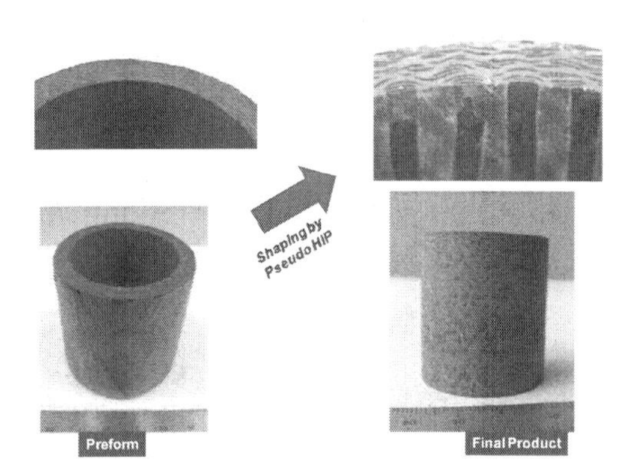

Fig. 8: "in-situ" Crystallized NITE-SiC/SiC Cylinder

Cast-ion and SiC/SiC hybrid component for the new die-cast machine is an example of metal/CMC hybrid materials. W wire embedded SiC/SiC heater is another example of hybrid material, where high temperature heating capability under neutron damage has been examined. Figure 9 provides W fiber embedded SiC/SiC heater and its comparison of electrical conductivity with W-3%Re and SiC/SiC. This material is providing attractive features as high temperature heater element for BR2 evaluation in Broader Approach between EU and Japan [5].

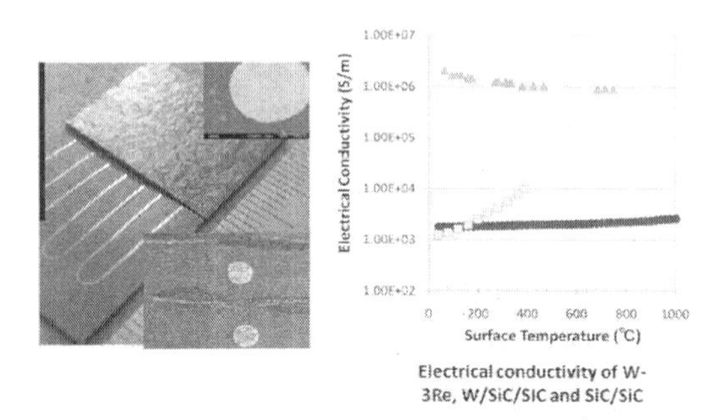

Fig. 9. W fiber embedded SiC/SiC heater component for reactor irradiation rig.

For aero-space technology development, SiC/SiC application to SST and HST had been performed as METI project, for fastener and combustion chamber liner. Those results were not promising, but current NITE-SiC/SiC rocket thrusters development is providing very promising results. As shown in Fig. 9, 500N class thrusters has been successful to make a high density preform which will be pseudo-hipped into the final products very soon.

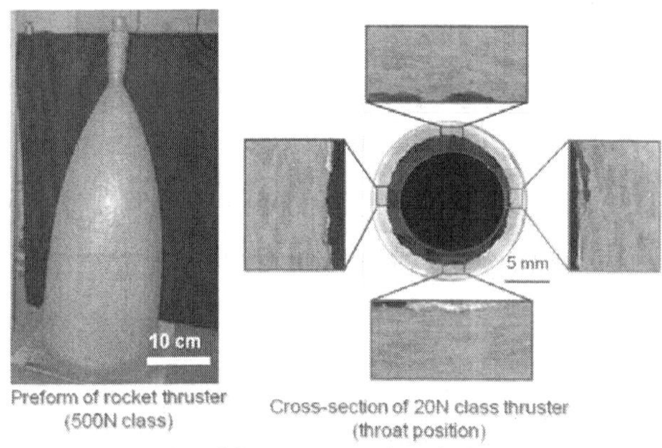

Already 20N class thrusters were made and the structure soundness is also shown in Fig. 10.

Fig. 10. All SiC/SiC Rocket Thrusters by NITE-method [6].

Step toward industrialization can be divided into 4 steps. The step starts from conceptual design step and through preliminary design step and detailed design step finally products design step becomes the goal. Figure 11 indicates the steps toward industrialization with the current status of NITE-SiC/SiC R & D. NITE-SiC/SiC production has already finished the first conceptual step and almost at the end of detailed design. Now OASIS activity is trying to launch the detailed design step with process optimization. Including Broader Approach for fusion energy R & D, test method standardization and design code establishment has been initiated. Also, a rocket thruster for launcher of satellite design is almost at the final stage and the proto type thrusters will be fabricated early 2011 for the test by the end of 2011.

	DESIGN	FABRICATION	Testing/ Evaluation	Status
1	Conceptual Design	Process Scoping	Shapes/Density	Completed
2	Preliminary Design	Process Screening	Microstructure/ Baseline Prop.	*In Progress*
3	Detailed Design	Process Optimization	Specification oriented evaluation	Launching
4	Product Design	Process Standardization	Product Verification Data	Preparing

Fig. 11: Steps Toward Industrialization
-where NITE is-

5. CONCLUSIONS

(1) By the intensive activities of OASIS/MIT, many progresses in NITE-SiC/SiC industrialization have been and being made.
(2) Other CMCs, like C/SiC, and hybrid materials with metals and FRPs are under development aiming at early supply into market.
(3) Biggest issues are to present economical competitiveness with other old/innovative materials and to present potentiality of stable and sufficient industrial basis. Of those, OAIS is trying to solve.
(4) Currently, STREAM project for Aerospace and Energy areas are the core of OASIS/FEEMA activity.

ACKNOWLEDGEMENT

Authors express their sincere appreciation to the members of OASIS, Kohno Laboratory and Kishimoto Laboratory, Muroran Institute of Technology for their kind support and collaborative works.

REFERENCES

[1] FEEMA Program Directory, 2009, Muroran Institute of Technology
[2] Advanced SiC/SiC Ceramic Composites, Editors A. Kohyama, M.Singh,, H.T. Lin and Y. Katoh, Ceramics Transactions vol.144(American Ceramic Society, 2002)
[3] A. Kohyama, "The Present Status of SiC/SiC R & D for Nuclear Application in Japan", in this proceedings.
[4] N. Nakazato, H. Kishimoto, J. S. Park, H. C. Jung, Y. Kohno and A. Kohyama, "Effects of Preform Densification on Near Net Shaping of NITE-SiC/SiC Composites forming", in this proceedings.
[5] T. Abe, H. Kishimoto, J. S. Park, H. C. Jung, and A. Kohyama, "SiC/SiC and W/SiC/SiC Composite Heater by NITE-method for IFMIF and Fission Reactor Irradiation Rigs", in this proceedings.
[6] in OASIS Annual Report for 2010, Muroran Institute of Technology, 2010.

EFFECTS OF TWO-STEP SINTERING ON DENSIFICATION AND PERFORMANCE OF NEAR-NET SHAPED NITE-SiC/SiC COMPOSITES

N. Nakazato[1], H. Kishimoto[2], K. Shimoda[3], Y. Kohno[2] and A. Kohyama[2, 3]

[1]Graduate School of Mechanical Systems and Materials Engineering, Muroran Institute of Technology, Muroran, Hokkaido, 050-8585, Japan

[2]College of Design and Manufacturing Technology, Muroran Institute of Technology, Muroran, Hokkaido, 050-8585, Japan

[3]OASIS, Muroran Institute of Technology, Muroran, Hokkaido, 050-8585, Japan

ABSTRACT

In the nano-infiltration and transient eutectic-phase (NITE) process, preform densification followed by hot-pressing has been confirmed to be an effective option for suppression of large volumetric shrinkage (~50 Vol%) during near-net shaping. This study tries to present the concept for fabrication of high performance NITE-SiC/SiC composites with preform densification process and the two-step sintering process. In particular, the effects of the two-step sintering process were investigated on densification behavior, microstructure development and mechanical property change. The two-step sintering process demonstrated improvement in matrix densification, enhanced composites' density and suppression of pores in the intra-fiber-bundles under the appropriate microstructure control. In the composites with the two-step sintering process (T_{s2}=1700 ^{o}C), enhanced sintering performance was exhibited, especially intra-fiber-bundles. The ultimate bending strength of the 2D-SiC/SiC composites fabricated was about 240 MPa with sufficient pseudo-ductility to fracture presented. On the other hands, in the composites with the two-step sintering process (T_{s2}=1800 ^{o}C), the ultimate bending strength was about 270 MPa with little pseudo-ductility to fracture presented. These results suggest the clear improvement in total performance by the two-step sintering process comparing with that by the single step sintering process.

INTRODUCTION

SiC/SiC composites are promising candidate structural materials for advanced nuclear energy systems due to their potentiality for providing excellent mechanical at high-temperature and low activation[1-3]. SiC/SiC composites by nano-infiltration and transient eutectic-phase (NITE) method present the great attractiveness[4-6]. One of the key fabrication methods for system components is the near-net shaping technique. In general, large volumetric shrinkage (~50 vol%) occurs during ceramic matrix composites fabrication by hot-pressing like NITE process due to infiltration and densification process of powders for matrix formation, resulting in significant fiber-architecture and -strength damages. Therefore, process development for suppression of large volumetric shrinkage during hot-pressing is indicated to be essential for damage-less near-net shaping, where preform densification before hot-pressing has been introduced as one of the potential option. The preform densification demonstrated the maintainability of fiber-architecture in complex shaped composites and the improvements of composites' relative density to 88 % and ultimate bending strength to ~200 MPa in 2D-SiC/SiC composites[7]. However, as the structural components, such as first wall and diverter of fusion reactors, the current composites' density and mechanical properties are still insufficient. In order to make additional improvement in fabricating high performance NITE-SiC/SiC composites, the two-step sintering process has been designed. This concept has been applied to fabrication of dense nano-structured monolithic ceramics (ex. Y_2O_3 or SiC)[8, 9]. The two-step sintering process is divided into; (1) transient liquid phase sintering process at the first-step temperature (T_{s1}) and (2) solid state sintering process at the second-step temperature (T_{s2}). In this study, sintering process based on the original NITE process was defined as "single-step sintering process".

OBJECTIVE

The objective of this study is to explore the fabrication of high performance NITE-SiC/SiC composites by the two-step sintering process followed to preform densification. In particular, the effects of two levels of second-step temperature (T_{s2}) of the two-step sintering process were investigated on densification behavior, microstructure development and mechanical property change.

EXPERIMENTAL PROCEDURE

Pyrocarbon (PyC) coated-Tyranno™ SA fibers (Ube Industrials Ltd., Japan) were used as reinforcement for SiC/SiC composites fabrications. The PyC coating was appropriately chosen at the thickness of 0.5 μm by chemical vapor deposition (CVD) process. β-SiC nano-powder with a mean grain size of 30 nm and sintering additives with Al_2O_3 (Kojundo Chemical Laboratory Co. Ltd., Japan, mean diameter of 0.3 μm, 99.99 %) and Y_2O_3 (Kojundo Chemical Laboratory Co. Ltd., Japan, mean diameter of 0.4 μm, 99.99 %) were used for matrix formation. For the fabrication of prepreg sheets, PyC-coated Tyranno-SA fibers were impregnated in 'nano'-slurry, which consisted of the mixture of β-SiC nano-powders and sintering additives. Prepreg sheets were stacked in $0^{o}/90^{o}$ cross-plied (CP) manner to prepare plate preforms. The preforms prepared were hot-pressed in Ar under a pressure of 10-20 MPa. T_{s1} and T_{s2} were 1800-1900 °C and 1700-1800 °C, respectively. The T_{s2} studied in this work are 1700 °C and 1800 °C. The composites hot-pressed were subsequently cut into 3 mm x 26 mm x 1.2 mm for three-point bending test. The three-point bending test was performed at room-temperature, with a crosshead displacement rate of 0.5 mm/min and outer support span of 16 mm, following ASTM C1341 general guidelines. The density of each sample was determined by

Archimedes method. Both of polished cross-sectional samples and the bending test sample fractures were observed by a field emission scanning electron microscopy (FE-SEM). The average grain size of SiC particles in matrix was measured by analysis of SEM images.

RESULTS AND DISCUSSION

Density and microstructure

Density and porosity of hot-pressed SiC/SiC composites are summarized in Table 1. Density is increased by the two-step sintering process. Figure 1 shows SEM images of cross section of SiC/SiC composites with the single-step and with the two-step sintering processes. No large pores could be identified at inter-fiber-bundles, as shown in Figure 1. Pores are mainly distributed in the intra-fiber-bundles regions. Porosity is decreased by the two-step sintering process, and it decreases with increasing T_{s2}. In the composites with T_{s2}=1800 °C, pores in the intra-fiber-bundles almost unobserved. Figure 2 shows SEM images of microstructure in inter- and intra-fiber-bundles matrix. In both of the composites with T_{s2}=1700 °C and 1800 °C, enhanced grain growth of SiC nano-particles observed. For the case of T_{s2}=1700 °C, the average grain size of matrix in the inter-fiber-bundles becomes to 1.4 times, and that of matrix in the intra-fiber-bundles also becomes to 1.9 times in comparison with the composites with the single-step sintering process. This result suggests that the

Table 1

Density and porosity of hot-pressed SiC/SiC composites with the single-step and with the two-step sintering processes.

	Density (g/cm^3)	Open prosity (%)	ED/TD[a] (%)	Fiber volume (%)
Single-step sintering process	2.8	7.4	88	40
Two-step sintering process (T_{s2} = 1700 °C)	2.9	4.0	92	41
Two-step sintering process (T_{s2} = 1800 °C)	3.0	3.0	96	42

[a] ED: experimental density, TD: theoretical density.

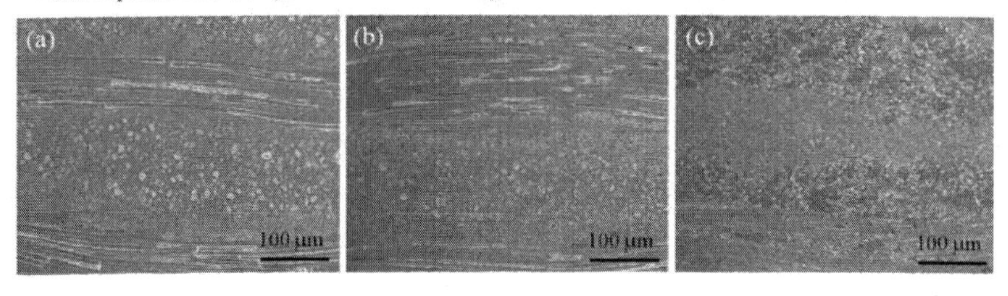

Figure 1: SEM images of cross section of SiC/SiC composites made: (a) with the single-step sintering process, (b) with T_{s2}=1700 °C, and (c) with T_{s2}=1800 °C.

two-step sintering process is effective to enhance sintering performance, especially intra-fiber-bundles. Both of the average grain size in inter- and intra-fiber-bundles of composites with T_{s2}=1800 °C reached to 1.3 times bigger than that of the composites with T_{s2}=1700 °C. Figure 3 shows SEM images of microstructure in the intra-fiber-bundles and fiber. Where, PyC interphase of SiC fibers have been kept in both of composites with the single-step sintering process and with T_{s2}=1700 °C. In these cases, there is no significant fiber deformation observed. For the case of composites with T_{s2}=1800 °C, deformation and grain growth of fibers and damage and loss of PyC interphase were clearly observed. The process improvement by the two-step sintering process was conceptually defined as, (1) the first-step for transient liquid phase sintering process and (2) the second-step for solid state sintering process. From the experimental results, the basic verification of the concept has been done. Furthermore, this two-step sintering process largely suppressed the damage to the PyC interphase and fiber which may bring further improvement of the composite performance by fine adjustment of the process condition to come.

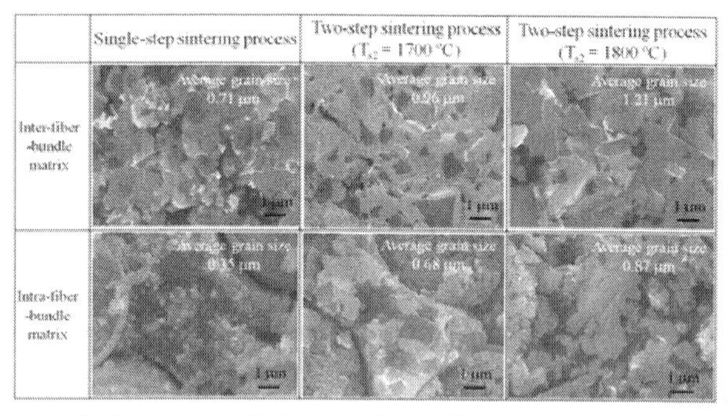

Figure 2: SEM images of microstructure in inter- and intra-fiber-bundles matrix of composites with the single-step and with the two-step sintering processes.

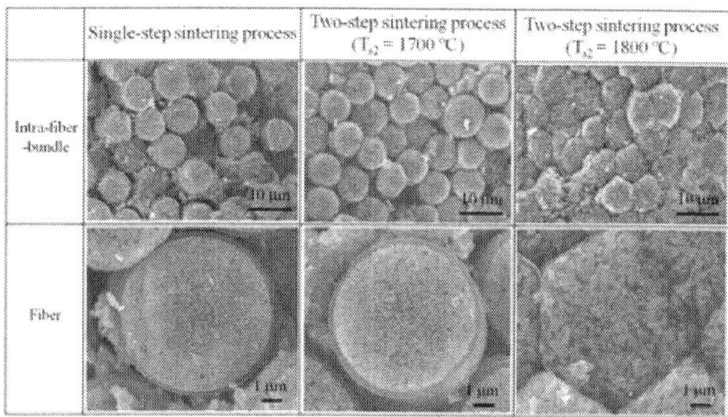

Figure 3: SEM images of microstructure in the intra-fiber-bundles and fiber of composites with the single-step and with the two-step sintering processes.

Mechanical properties and fracture behavior

Figure 4 shows flexural test results of the composites studied. The sample number for each composite was three. The average ultimate bending strength (UBS) of the composites with T_{s2}=1700 °C was about 240 MPa, which reached to 1.2 times higher than that of the composites with the single-step sintering process. In addition, the composites with the single-step sintering process and with T_{s2}=1700 °C displayed a pseudo-ductility fracture behavior. Whereas, for the case of composites with T_{s2}=1800 °C, the average UBS was about 270 MPa, which reached to 1.4 times higher than that of the composites with the single-step sintering process and the fracture mode was typical brittle fracture. Figure 5 shows typical fracture surface of the composites with the single-step and with the two-step sintering process after the three-point bending test. Fiber pull-outs were observed in Figure 5 (a) and (b) and no fiber pull-outs seen in figure 5 (c). The increasing of UBS of the composites with the two-step sintering processes might be due to matrix densification, especially intra-fiber-bundles. On the other hands, brittle fracture behavior of the composites with T_{s2}=1800 °C seems to be due to fiber degradation and damaged and loss of PyC interphase.

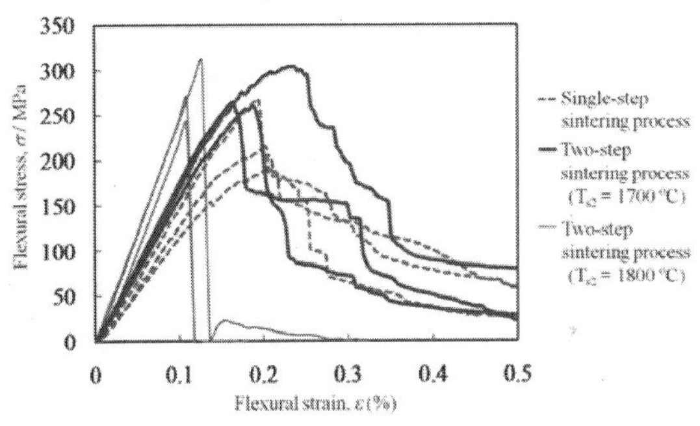

Figure 4: Flexural stress-strain curves of composites with the single-step and with the two-step sintering processes.

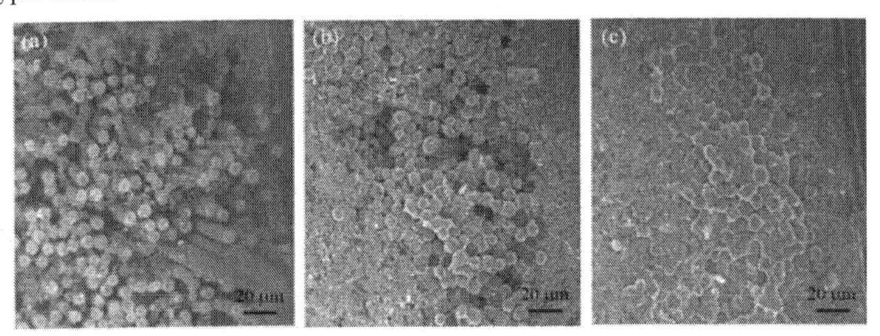

Figure 5: Fracture surface of SiC/SiC composites with the single-step and with the two-step sintering processes: (a) with the single-step sintering process, (b) with T_{s2}=1700 °C, (c) with T_{s2}=1800 °C.

CONCLUSION

For the further improvement in performance of NITE-SiC/SiC composites, the concept verification of the two-step sintering process was studied and the preliminary results indicated the potential attractiveness of the two-step sintering process was confirmed in addition with the preform densification followed by the two-step sintering process. The two-step sintering process demonstrated enhanced composites' density and suppression of pores in the intra-fiber-bundles. But, in this process evaluation, composites with T_{s2}=1800 °C looks too high, causing large degradation of pseudo-ductility fracture behavior might be due to damaging of PyC interphase and too much grain growth and fiber degradation. For the case of composites with T_{s2}=1700 °C, pseudo-ductility was kept and a slight strength improvement was observed. However, it was too unclear to conclude the second-step temperature (T_{s2}) modification direction toward the process optimization. In the further work, further elementary process analysis of the two-step sintering process is to be done.

REFERENCES

[1]R. Naslain, Design, Preparation and properties of non-oxide CMCs for application in engines and nuclear reactors: an overview, *Compos. Sci. Technol.*, **64**, 155-170 (2004)

[2]A. Kohyama, M. Seki, K. Abe, T. Muroga, H. Matsui, S. Jitsukawa and S. Matsuda, Interactions between fusion materials R&D and other technologies, *J. Nucl. Mater.*, **283-287**, 20-27 (2000)

[3]L. L. Snead, R. Jones, A. Kohyama and P. Status, Status of silicon carbide composites for fusion, *J. Nucl. Mater.*, **233-237**, 26-36 (1996)

[4]A. Kohyama, S. M. Dong and Y. Katoh, Development of SiC/SiC composites by Nano-Infiltration and Transient Eutectoid (NITE) process, *Ceram. Eng. Sci. Proc.*, **23**, 311-318 (2002)

[5]Y. Katoh, S. M. Dong and A. Kohyama, Thermo-mechanical properties and microstructure of silicon carbide composites fabricated by nano-infiltrated transient eutectoid process, *Fusion Eng. Des.*, **61-62**, 723-731 (2002)

[6]S. M. Dong, Y. Katoh and A. Kohyama, Preparation of SiC/SiC composites by hot pressing, using Tyranno-SA fiber as reinforcement, *J. Am. Ceram. Soc.*, **86**, 26-32 (2003)

[7]N. Nakazato, H. Kishimoto, K. Shimoda, J. S. Park, H. C. Jung, Y. Kohno and A. Kohyama, Effects of preform densification on near-net shaping of NITE-SiC/SiC composites, Proceedings of ICC3, in printing

[8]I. W. Chen and X. H. Wang, Sintering dense nanocrystalline ceramics without final-stage grain growth, *Nature*, **404**, 168-171 (2000)

[9]Y. I. Lee, Y. W. Kim, M. Mitomo and D. Y. Kim, Fabrication of Dense Nanostructured Silicon Carbide Ceramics through Two-Step Sintering, *J. Am. Ceram. Soc.*, **86**, 1803-1805 (2003)

Ceramics for Electric Energy Generation, Storage, and Distribution

CERAMIC PROCESSING FOR DENSE MAGNESIUM DIBORIDE

C E J Dancer*, R I Todd and C R M Grovenor
Department of Materials, University of Oxford, Parks Road, Oxford, OX1 3PH, UK

*Corresponding author: claire.dancer@materials.ox.ac.uk

ABSTRACT

There is little consensus in the literature on the processing requirements to produce high quality magnesium diboride (MgB_2) material with low impurity content and high density. We have carried out a range of processing procedures to establish key parameters for ex situ production of dense MgB_2 by examining the effect of modifying the characteristics of the starting powder and the processing parameters during heat-treatment. MgB_2 bulk material was produced from as-purchased and modified powders by pressureless heat-treatment under Ar gas and pressure-assisted methods (spark plasma sintering or hot pressing) under vacuum or Ar gas. Porosity, impurity content, hardness, grain size and bulk superconducting properties were measured. Our results indicate that densification and applied pressure are strongly correlated, while the effect of temperature is less significant. The optimum processing environment (inert gas or vacuum) was dependent on the technique used. These results indicate that pressure-assisted heat-treatment is required in order to produce dense bulk MgB_2. A positive correlation between magnetization critical current density and bulk density was observed for magnesium diboride bulks of up to around 90% density. Above this level, other microstructural processes such as grain growth begin to influence the critical current density, suggesting that full elimination of porosity is not necessary to obtain good superconducting performance.

INTRODUCTION

Magnesium diboride (MgB_2) superconductors have been widely produced in the form of powder-in-tube (PIT) wires and tapes since very shortly after the first observation of superconductivity in MgB_2 by Nagamatsu et al[1]. Processing routes to produce these PIT wires and tapes can be classified into two types: in situ and ex situ. To produce PIT wires by in situ processing techniques, precursor materials (such as magnesium and boron powders) are filled into metal tubes which are plugged and swaged, drawn, and/or rolled. These wires are then heated to the required temperature to form the superconducting MgB_2 phase. While this enables the easy addition of beneficial impurities such as carbon, the large density change on conversion of precursors to the MgB_2 phase means that the PIT wires are difficult to make fully dense[2]. This means that a significant proportion of the volume of the PIT wire or tape will be non-superconducting, which limits the attainable critical current density.

In contrast, the less widely-studied ex situ route uses pre-reacted MgB_2 powder, such that the PIT wires can superconduct without any heat-treatment[3]. However it has generally been found that heat treatment at 900-1000°C is beneficial for superconducting properties without resulting in the formation of extensive detrimental reaction layers with the sheath materials[4]. Because the MgB_2 phase is present from the start, and because densification by sintering is minimal at 1000°C, no change in density occurs during heat-treatment. For this reason, ex situ processing is preferred to in situ processing by at least one major commercial MgB_2 wire manufacturer[5].

While a positive correlation between sample bulk density and superconducting performance has previously been established[6], relatively little research has been carried out on the processing requirements to produce dense magnesium diboride by the ex situ method for use in PIT wires and tapes. In this paper, we summarise the findings of a research project[7-10] which was designed to establish the conditions required to produce dense MgB_2 bulk material using a selection of standard ceramic processing techniques and a standard, widely-used commercially produced MgB_2 powder.

PROPERTIES OF BULK MgB_2 PRODUCED FROM AS-SUPPLIED POWDER

MgB_2 bulks were produced from as-purchased MgB_2 powder (Alfa Aesar, 98% purity, -325 mesh). Characterisation of these samples provided a reference point for the microstructural and superconducting properties against which the effect of later changes to the processing could be assessed. For the same reason, the purity and particle size of the MgB_2 powder was measured using X-ray diffraction (XRD; Philips θ-2θ diffractometer with Cu Kα radiation generated at 35 kV and 50 mA) and laser diffraction (Malvern Mastersizer S with propan-2-ol). The particular batch of MgB_2 powder used in this work had an XRD spectrum containing only peaks attributable to the MgB_2 phase. As noted by other researchers using this powder[11,12], the particle size distribution was extremely broad, with a bimodal distribution containing particles of 0.5-200 μm when measured by laser diffraction without any dispersion (described as "dry" measurements in this work) (Figure 1). However, unlike earlier studies, we found that when the powder was dispersed in propan-2-ol by ultrasonication for 300 s, the particle size distribution shifted to a lower range (0.2-100 μm) and the number of small particles increased. This shows that the as-purchased MgB_2 powder contains a high proportion of agglomerated particles, and that the true particle size range is significantly lower than that previously reported.

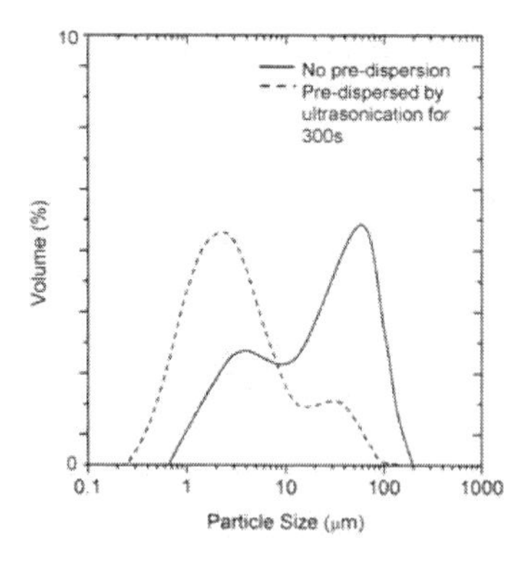

Figure 1: Particle size distribution in as-purchased MgB_2 powder. Powder was either added dry to the propan-2-ol in the Mastersizer system with no pre-dispersion, or was pre-dispersed by ultrasonication for 300 s in propan-2-ol, then added to the system as slurry.

Bulk MgB_2 samples were produced by pressing 0.5 g of MgB_2 powder into pellets of approximately 9 mm diameter and 4 mm thickness. To avoid loss of magnesium during the heat-treatment, the pellets were placed inside an MgB_2 powder bed within a closed alumina crucible. These were then heat-treated for 1 h in flowing argon gas at a peak temperature of between 200 and 1100 °C. Temperatures above this level would result in significant oxidation of the material, and in addition are impractical for the production of PIT wires and tapes. However, this pressureless heat-treatment of MgB_2 at temperatures up to 1100 °C resulted in very limited sintering. The pellets were characterised by density measurements by the Archimedes method using propan-2-ol as the immersion medium, by Vickers hardness measurements (room temperature, 2 kg load, 10 s dwell time), by XRD and by scanning electron microscopy (SEM). An increase in hardness was observed after heat-treatment at all temperatures[8], but this was not accompanied by a change in density, indicating that sintering was in its very early stages. The hardness increased with heat-treatment temperature, indicating an increased connectivity between particles. However no difference in the degree of connectivity between grains could be observed by SEM on fracture surfaces of the samples compared to that before heat-treatment (Figure 2a and 2b).

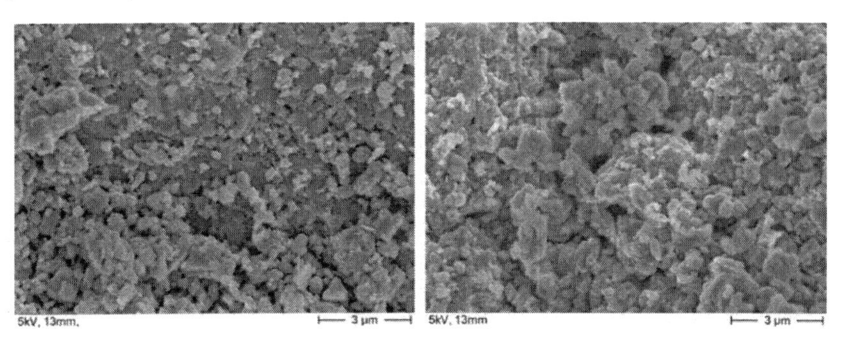

Figure 2: SEM micrographs showing the fracture surfaces of pressureless heat-treated MgB_2 bulks (left) as-pressed (right) after heat-treatment at 950°C. No increase in connectivity between particles was observed.

Magnetization measurements of selected samples produced by pressureless heat-treatment were measured using a Quantum Design Magnetic Property Measurement System (MPMS). Measurements were made at 4 and 20 K with an applied magnetic field of up to 7 T, with a smaller step size of 0.05 T for $-2 \leq H(T) \leq 2$ and a larger step size of 0.2 T for $2 \leq H(T) \leq 7$ and $-7 \leq H(T) \leq -2$. The critical current density ($J_{c,mag}$) was calculated using the Bean model[13]. For all samples examined, a higher $J_{c,mag}$ of up to 8×10^4 A cm^{-2} at 0 T was measured at low magnetic fields, but poor flux pinning and connectivity meant $J_{c,mag}$ decreased rapidly with increasing applied magnetic field.

EFFECT OF MODIFICATION OF STARTING POWDER

The particle size distribution of the as-purchased MgB_2 powder was modified by ball milling and attrition milling. Ball milling was carried out in sealed 250 ml polyethylene jars with propan-2-ol as the dispersion medium and using yttria-stabilized zirconia (YSZ) milling media (Inframat Advanced Materials, cylindrical, 10 mm diameter, 10 mm height). Milling was carried out for durations of 8 and 20 h. Attrition milling was carried out using a commercially available attrition mill (Union Process Szegardi Attritor system, model 01HD) with a YSZ tank and blade and spherical YSZ milling media (Inframat Advanced Materials, 3mm diameter beads). Attrition milling was carried out for 2 h at 500 rpm using propan-2-ol as the dispersant.

XRD spectra taken from the modified powders showed that, while the ball milled powders remained unchanged, some contamination by ZrO_2 occurred during attrition milling as one peak attributable to ZrO_2 was observed (Figure 3). By comparison to the work of Chen et al[14], who deliberately added ZrO_2 additions to MgB_2 ex situ and in situ powders, the XRD peak intensity for ZrO_2 is extremely low. It was therefore not possible to accurately assess the ZrO_2 content from the XRD data. Elemental XPS analysis showed that the Zr content of the surfaces of the attrition milled powder was 0.6 at% indicating that the total ZrO_2 content of the powder is expected to be significantly lower than this value. In addition, laser diffraction measurements showed that the powder is sensitive to exposure to dispersing liquids, and re-agglomerates readily during drying. This could be seen by measuring the particle size without pre-dispersing the powder in the propan-2-ol. The powders again had significantly different particle size distributions following dispersion by ultrasonication for 300 s

(Figure 4). Of particular note is the attrition milled powder, which appears from the "dry" measurements to contain mostly large particles. Ultrasonication reveals that these are in fact agglomerates, and that a large proportion of sub-1 μm particles are present.

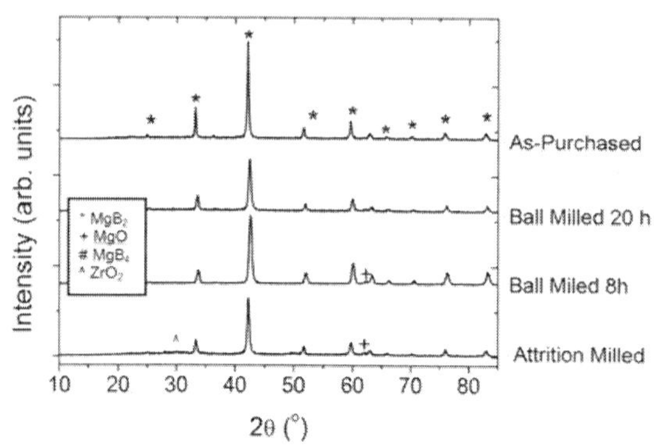

Figure 3: X-ray diffraction spectra of modified MgB_2 powders. The XRD trace of the as-purchased MgB_2 powder is included for comparison.

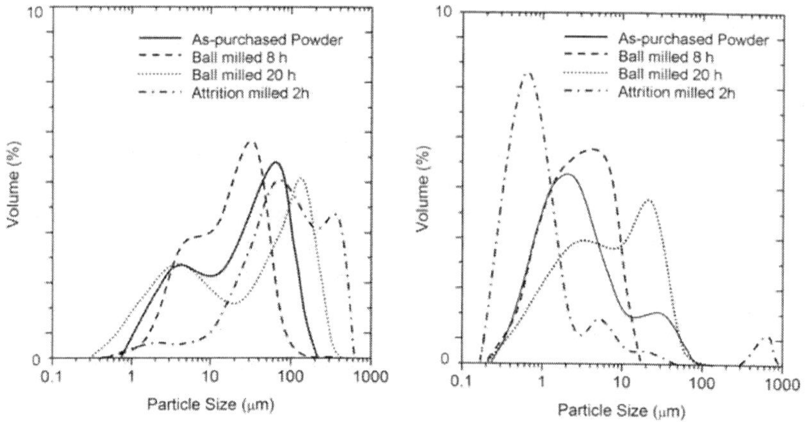

Figure 4: Particle size distributions of milled powders, measured by laser diffraction. The equivalent measurements for as-purchased powder are included for comparison. (Left) no dispersion. (Right) dispersed by ultrasonication.

Bulk MgB_2 samples were produced from all milled powders by the same route described above for the as-purchased powder, and were heat-treated for 1 h at peak temperatures of 950 and 1100°C. These were then characterised in the same ways described earlier. XRD analysis (Figure 5) showed that many of the bulk samples produced from modified powders had relatively high impurity content compared to the starting powder. MgO was the most significant impurity phase formed, but MgB_4 was also seen in several samples. Greater proportions of impurity phases formed for samples heat-treated at higher temperatures. The relative densities of the samples were calculated by reference to the theoretical density calculated from the densities of the individual components using a rule of mixtures and the proportions of MgO and MgB_2 measured from XRD spectra using a calibration curve[8].

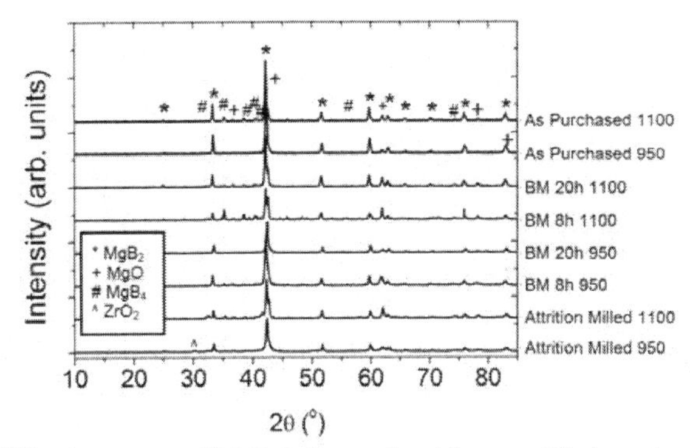

Figure 5: X-ray diffraction spectra of MgB_2 bulks produced from modified powders. The XRD traces of the bulks produced from as-purchased MgB_2 powder are included for comparison.

In most cases the densities of the samples produced from milled powders were lower (51-61%) than those produced from as-purchased powder (65-70%). However, despite this, there are reasons to expect that some increase in connectivity between grains has occurred by examining the measured magnetization data (Figure 6) for selected samples pressureless heat-treated at 1100°C. The $J_{c,mag}$ of the sample from the powder produced by ball milling for 20 h was lower than that of all the other samples, including that produced from as-purchased powder. This contradicts the findings of Braccini et al[15], who found that J_c increases with milling time. We believe this difference is due to the use of different milling procedures, and also the effect of the time of exposure to propan-2-ol, which, as discussed earlier, has a significant effect on the agglomeration and impurity content of the milled powders. In low magnetic fields the sample produced from attrition milled powder had similar performance to that from the as-purchased powder, but in high fields the $J_{c,mag}$ values were higher by one order of magnitude. As discussed earlier, the attrition powder was the only one to show any detectable ZrO_2 contamination from the milling media, so it is possible that the improved performance is due to the ZrO_2 addition. However Chen et al[14] found that ZrO_2 addition depresses superconductivity in MgB_2 bulks, though it should be noted that their results are for samples produced by the in situ method. The sample produced from powder which had been ball milled for 8 h has the best performance in low field, though that from attrition milled powder is superior in high fields.

From these results it is apparent that modifying MgB_2 powder by milling processes changes the superconducting properties of the material and under certain conditions results in an improvement over the as-purchased powder. However the milling processes used here did not result in an increase to the bulk density of the samples because (a) the agglomerated particles are inherently porous and (b) agglomeration increased due to the exposure to the dispersing liquid. Careful control over the milling conditions is therefore required to maximise the performance of MgB_2 bulk material produced in this way.

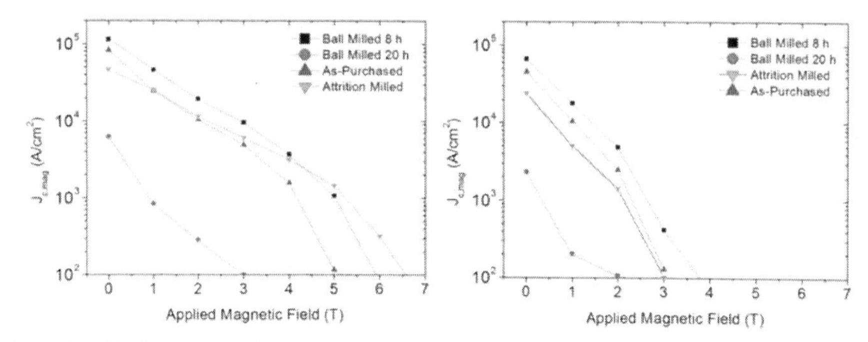

Figure 6: Critical current density measurements for selected samples produced from modified powders, and compared to the results for as-purchased powder. Measurements were taken at temperatures of 4 K (left) and 20 K (right).

EFFECT OF PRESSURE-ASSISTED HEAT-TREATMENT

Pressure-assisted heat-treatment methods such as hot pressing or spark plasma sintering (SPS) involve the application of pressure while the powder compact is heated to high temperatures. The use of pressure during heat-treatment is known to produce dense, fine grained material in ceramics such as alumina[16], and can reduce the temperature required to sinter to full density, and has been previously used to produce dense MgB_2 bulks using a fixed pressure[17]. To establish the optimum processing parameters for MgB_2 we also investigated the effect of the variation of pressure on the densification of MgB_2 by SPS.

SPS was carried out using a SPS furnace (Model HPD25/1, manufactured by FCT Systeme GmbH, Germany) based at Queen Mary, University of London. In SPS, the sample is heated very rapidly by direct current applied across the die. As-purchased MgB_2 powder was placed in a graphite die lined with graphite paper and spark plasma sintered under vacuum to a maximum temperature of 1250°C and 16, 50 or 80 MPa uniaxial pressure. The samples were removed from the die and the surfaces ground using a flat bed grinder to remove carbon-contaminated outer layers. The samples were then characterised by the same methods as were described earlier for pressure-less heat-treated samples.

Significant densification occurred during the SPS process. The die ram height position, die temperature, and applied pressure (not shown) were recorded during the cycle, and it is evident that a significant decrease in the height of the sample occurred above 1000°C (Figure 7). The applied pressure is at its maximum value well before the time at which the die reaches this temperature. Further examples of these data for other pressures have been published previously[9].

This sudden height change, beginning at just over 1000°C, was observed in all samples, and is therefore taken to be the onset sintering temperature for MgB_2. This broadly agrees with the work of Tampeiri et al[18] who found that sintering in MgB_2 during hot pressing under 30 MPa uniaxial pressure began at a temperature of 1050°C, and of that of Shim et al[17], who found a temperature of 1000-1050°C is suitable for SPS processing of MgB_2 in argon gas. The two samples sintered at 50 and 80 MPa uniaxial pressure had very high densities of 2.63 g cm^{-3} while that sintered at 16 MPa had a bulk density of 2.52 g cm^{-3}. The theoretical maximum density of pure MgB_2 is 2.63 g cm^{-3}. However these samples all contain around 8 wt.% MgO impurity phase measured by XRD. When this is accounted for, the relative densities for these samples were calculated as 97% for those processed using 50 and 80 MPa uniaxial pressure, and 92% for that processed using 16 MPa. The densities of all of these samples are significantly higher by c.15-20% compared to MgB_2 bulks produced by Locci et al[19] by SPS at 50 MPa pressure using the in situ route with Mg and B starting powders. For the ex situ route, Shim et al[17] report relative densities of >99%; however while they state that MgO was found in the SPSed samples, and show XRD data confirming this, they do not state the MgO content or whether they have taken this into account in calculating the relative bulk densities, therefore it is not possible to compare their results to our own. Takano et al[20], who estimate the MgO content at 15% from SEM images, measure the relative density of their ex situ MgB_2 bulks prepared by HIPing at 3.5 GPa to be 2.66 g/cm^3. Taking into account the 15% MgO content, we calculate the density of their samples as 96%, which is similar to that achieved using SPS and a significantly lower pressure. Observations of these samples using SEM agreed with our bulk density measurements (Figure 8). The fracture surfaces of samples pressed at 50 and 80 MPa consisting of large dense regions with a few more porous areas. By comparison, the sample pressed at 16 MPa had significantly more porosity.

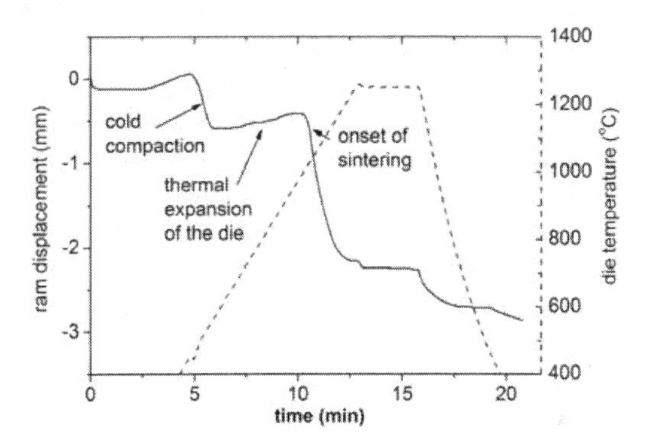

Figure 7: An example of data recorded during an SPS cycle, showing the onset of sintering as the die temperature reaches 1000°C. The solid line shows the ram displacement while the dashed line shows the die temperature.

The magnetization data (Figure 9) for these samples clearly show an improvement in the critical current density at all magnetic fields for those samples which were processed by SPS compared to the pressureless heat-treated samples. This is indicative of increased connectivity in the material, as demonstrated by Liao et al[21] by their work on hot isostatic pressing of MgB_2 wires. However, it is noticeable that there is little improvement in critical current density for the increase in uniaxial pressure from 16 MPa to 80 MPa, i.e. the increase in density from 92 to 97% of the maximum. This indicates that complete densification may not be critical for the achievement of good superconducting properties. Above c.90% density the grain size would be expected to increase, which would reduce the critical current density. It is therefore possible that a balance between the effects of porosity and fine grain size on the superconducting properties means that the optimal density may actually be somewhat less than 100%. Further work is needed to test this hypothesis, as there is surprisingly little comparable data in the literature for $J_{c,mag}$ measurements of MgB_2 bulks made by the ex situ method, and comparisons to the in situ method are unadvisable due to the density change on the conversion of Mg + B powders to MgB_2, which inevitably leads to lower densities for materials processed via the in situ route[19]. Hsieh et al[22] obtain almost identical $J_{c,mag}$ values to our samples for hot pressed MgB_2, though the highest J_c values for MgB_2 bulks are around $1x10^6$ A cm^{-2} for samples produced by hot isostatic pressing at pressures of 100-200 MPa [20,23]. The higher J_c values may be due to enhanced pinning due to more favourable defect or impurity distribution, or the suppression of grain growth by the higher pressure than can be achieved by SPS or hot pressing. However, the microstructural data needed to test this hypothesis is not available in the literature.

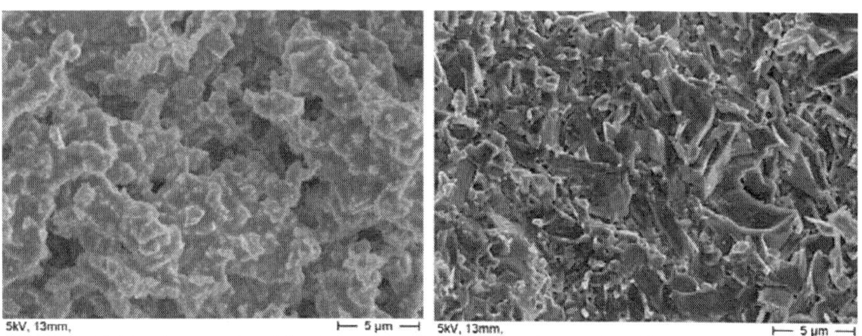

Figure 8: SEM micrographs of the fracture surfaces of samples produced by SPS. Samples were processed under uniaxial pressure of 16 MPa (left) and 50 MPa (right).

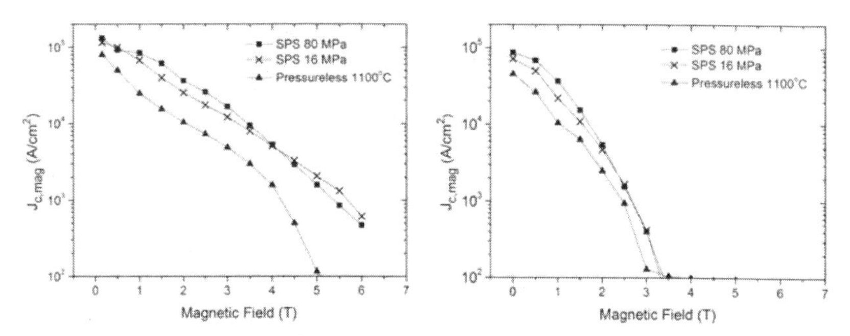

Figure 9: Critical current density of samples processed by SPS under uniaxial pressure of 16 and 80 MPa, compared to data from a sample produced from the same powder by pressureless heat-treatment. Measurements were taken at temperatures of 4 K (left) and 20 K (right)

To demonstrate the importance of applied pressure to the density of MgB_2, a plot of pressure versus both hardness and density for all samples studied in this work has been plotted and compared to a similar plot of temperature versus hardness and density (Figure 10).

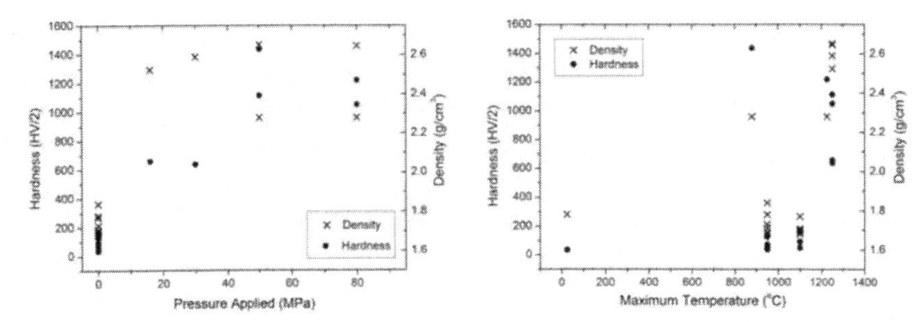

Figure 10: The effects of uniaxial pressure (left) and heat-treatment temperature (right) on density and hardness of MgB_2 samples.

From these data it is clear that the application of high pressure during heat-treatment leads to both high hardness and density, whereas the plot for temperature shows no correlation (samples with very high MgO content have been excluded from these analyses). Shim et al[17] prepared MgB_2 bulks by by SPS and temperatures of 950-1050°C, however they used a fixed pressure of 50 MPa so their data do not reveal the effect of pressure on the density of MgB_2. Our work indicates that the application of high pressure during heat-treatment is critical and leads to highly dense MgB_2 bulk material with a small grain size, which is important for obtaining good superconducting properties as the high concentration of grain boundaries increases the flux pinning in high applied magnetic fields[24,25]. Hot pressing causes an increase in the driving force for densification in ceramics[16]. This means that dense material can rapidly be obtained. In the case of MgB_2, a rapid heat-treatment at as low a temperature as possible is crucial in order to prevent the formation of excessive impurity content, as magnesium is very volatile and reactive.

CONCLUSIONS

As-purchased Alfa Aesar MgB_2 powder is highly agglomerated, and the measurement of particle size depends strongly on the degree of dispersion carried out prior to examination. The powder can be de-agglomerated using ultrasound to give a reasonably fine size distribution. The modification of the particle size distribution of MgB_2 powder can be achieved by ball milling or attrition milling in propan-2-ol for relatively short time periods. However, the drying of MgB_2 powders by evaporation following long exposure to propan-2-ol leads to the formation of harder agglomerates than in the starting powder, which could not be easily broken up by ultrasonication. These agglomerates prevented significant improvements in bulk density. Short duration milling procedures resulted in the most improvement in critical current density in both high and low magnetic fields.

The application of pressure of as little as 16 MPa during heat-treatment using the SPS technique above 1000°C was found to result in a significant improvement in both the bulk density and the critical current density compared to samples produced from the same powder by pressureless heat-treatment. Relative densities of 97% were achieved for applied uniaxial pressure above 50 MPa, which is similar to results obtained by hot isostatic pressing at significantly higher pressures. However little difference was observed in the critical current densities of samples produced at different uniaxial pressures, suggesting that complete densification is not necessarily required to achieve improved superconducting properties.

ACKNOWLEDGEMENTS

This work was funded by the EPSRC via a studentship from the Doctoral Training Account of the Department of Materials, University of Oxford. The authors would like to thank the following for their assistance – Dr Pavel Mikheenko, Dr Alex Bevan and Prof. Stuart Abell of the University of Birmingham; Dr Mike Reece and Dr Haixue Yan of Queen Mary, University of London; Prof. Ekrem Yanmaz and Mehmet Başoğlu of Karadeniz Technical University, Turkey; Dr Dharmalingham Prabhakaran of the Department of Physics, University of Oxford; Dr Alison Crossley, Frank Cullen and Chris Salter of the Department of Materials, University of Oxford.

REFERENCES

1. Nagamatsu J, Nakagawa N, Muranaka T, Zenitani Y and Akimitsu J (2001) Nature v.410 pp. 63-64
2. Zhou S, Pan A V, Liu H and Dou S (2002) Physica C v.382 pp.349-354
3. Grasso G, Malagoli A, Ferdeghini C, Roncallo S, Braccini V, Siri A S and Cimberle M R (2001) App. Phys. Lett. v.79 pp. 230-232
4. Kováč P, Hušek I and Melišek T (2002) Supercond. Sci. Technol. v.15 pp.1340-1344
5. Penco R and Grasso G (2007) IEEE Trans. Appl. Supercond. v.17 pp.2291-2294
6. Grinenko V, Krasnoperov E P, Stoliarov V A, Bush A A and Mikhajlov B P (2006) Solid State Commun. v.138 pp.461-465
7. Dancer C E J, DPhil Thesis, University of Oxford, (2008)
8. Dancer C E J, Mikheenko P, Bevan A, Abell J S, Todd R I and Grovenor C R M (2008) J. Eur. Ceram. Soc v.29 pp.1817-1824
9. Dancer C E J, Prabhhakaran D, Başoğlu M, Yanmaz E, Yan H, Reece M, Todd R I and Grovenor C R M (2009) Supercond. Sci. Technol. v.22 095003
10. Dancer C E J, , Prabhhakaran D, Crossley A, Todd R I and Grovenor C R M (2010) Supercond. Sci. Technol. v.23 065015
11. Flukiger R, Lezza P, Beneduce C, Musolino N, and Suo H L (2003) Supercond. Sci. Technol. v. 16 pp. 264-270
12. Goldacker, W, Schlachter S I, Liu B, Obst B and Klimenko E (2004) Physica C v.401 p.80-86
13. Bean C P (1964) Rev. Mod. Phys. v.36, pp.31-39
14. Chen S K, Glowacki B A, MacManus-Driscoll J L, Vickers M E and Majoros M (2004) Supercond. Sci. Technol. v. 17 pp. 243-248
15. Braccini V, Nardelli D, Penco R and Grasso G (2007) Physica C v.456 pp.209
16. Reed J S Principles of Ceramic Processing. 2nd ed. (1995) John Wiley and Sons, New York
17. Shim S, Shim K and Yoon J (2005) J. Am. Ceram. Soc. v.88 pp.858-861
18. Tampieri A, Celotti G, Sprio S, Caciuffo R and Rinaldi D (2004) Physica C v.400 pp.97-104
19. Locci A M, Orrù R, Cao G, Sanna S, Congiu F, and Concas G (2006) AIChE Journal v.52 pp.2618-2626
20. Takano Y, Takeya H, Fujii H, Kumakura H, Hatano T, Togano K, Kito H and Ihara H (2001) Appl. Phys. Lett. v.78 pp. 2914-2916
21. Liao X Z, Serquis A, Zhu Y T, Civale L, Hammon D L, Peterson D E, Mueller F M, Nesterenko V F and Gu Y (2003) Supercond. Sci. Technol. v.16 pp.799-803
22. Hsieh C H, Chang C H, Chang C N, Sou U C, Sheu H S, Hsu H C and Yang H C (2006) Solid State Commun. v.137 pp.97-100
23. Shields T C, Kawano K, Holdom D and Abell J S (2002) Supercond. Sci. Technol. v.15 pp.202–205
24. Toulemonde P, Musolino N and Flukiger R (2003) Supercond. Sci. Technol. v.16 pp.231-236
25. Cimberle M R, Novak M, Manfrinetti P and Palenzona A (2002) Supercond. Sci. Technol. v.15 pp.43-47

INVESTIGATION ON PHASE TRANSFORMATION OF YBCO-IN_2O_3 COMPOSITE SUPERCONDUCTOR COOLED DOWN VIA DIFFERENT ROUTES

Ahmed Y.M.Z.*[a], Hassan M.S.*, Abd-Elatif H.**

* Central Metallurgical Research and Development Institute, CMRDI
P.O.Box: 87 Helwan, El-Felzat St, El-Tebin, Cairo, Egypt
** Higher Institute of Optics Technology
60 Obour St., Misr El-Gedida, Cairo, Egypt
e-mail: ahmedymz@hotmail.com

ABSTRACT

Samples of the composite superconducting system YBCO-In_2O_3 were prepared and investigated using XRD, SEM, EDX and Temperature-resistivity measurements. The samples were divided into two identical classes which were cooled down from sintering temperature 940°C to room temperature by two different routes: 1- rapid quenching and 2- slow cooling (1- 2 °C/min.). XRD analysis showed that the addition of In_2O_3 caused a rapid phase transformation from tetragonal to orthorhombic in a very short time (45 seconds) upon quenching to room temperature. In all quenched doped samples, a phase difference between the sample surface (Orthorhombic) and its interior (Tetragonal) was observed while the slowly cooled samples didn't show such a phase difference. All slowly cooled samples showed a single orthorhombic phase. Doping the material YBCO with In_2O_3 led to the formation of a secondary phase that could be identified. This phase co-exists with the 123 parent phase without affecting its crystal structure and caused a gradual slight decrease in Tc-values with increasing In_2O_3 content. The bulk densities of the samples and their resistivity to corrosion in water were found to increase with increasing In_2O_3 content.

1. INTRODUCTION

In 1991, Paulose, K. V. et al [1] discovered that the pure superconducting phase YBCO doped with small amounts of Sb_2O_3 showed a rapid tetragonal to orthorhombic phase transformation by direct quenching from 950 °C to room temperature in flowing oxygen gas. By this experimental result they could prove that the widely accepted procedure of slow cooling of the samples (or its annealing for several hours at a constant temperature between 500- 600 °C) was not an essential condition to obtain the superconducting phase.

It is well known that the rates of oxygen absorption in pure YBCO phase are low at temperatures below 700 °C which makes the slow cooling in flowing oxygen gas an essential condition to obtain the orthorhombic superconducting phase [2-4]. However, there is an established experimental fact that a pure YBCO sample which is quenched in air or in oxygen from a temperature above 700 °C to room temperature is always found to have the tetragonal (semi-conducting) phase and never showed the 123 orthorhombic phase [2,5-7]. Accordingly, the results of Paulose experiment seemed abnormal and attracted the attention of several research groups who repeated and confirmed this observation. They also found that this phenomenon was working with the using of differnt dopant oxides like; Nb_2O_5, Ta_2O_5, WO_3, ZrO_2, and HfO_2 [8-14]. The role of the dopant oxide in the material remained unclear and until now this anomalous phase transformation has no concrete explanation.

Herein, the effect of addition various amount of In_2O_3 on the properties of YBCO samples were investigated. Ddifferent routes for cooling the sintered samples to room temperature in terms of slow cooling and quenching were applied and its effect on the final sample properties were studied. Both structure and superconducting properties of the samples prepared by the two cooling procedures were investigated. The bulk densities of all the samples were evaluated and their degradation in moisture was studied.

2. EXPERIMENTAL

Samples of the pure phase YBCO were prepared using solid-state reaction technique by mixing proper weights of the high-purity (99.99 %) Y_2O_3 (Sigma-Aldrich, Germany), and high purity 99.9 of both $BaCO_3$ and CuO (Spectrum chemical MFG. Corp., New Brunswick, USA). The resulting YBCO material was divided into two groups of samples. High purity (99.99 %) In_2O_3 (Sigma-Aldrich, Germany) was added to the samples of each of the two groups in small ratios x = 0.0, 0.3, 0.6, 0.9, 1.5, 2, 4, 5 and 7 % by weight to prepare the composite system $[YBa_2Cu_3O_{7-y}]_{1-x}(In_2O_3)_x$. All the samples were grounded and compressed into pellets of thickness 6 mm and diameter 14 mm. Then the samples were sintered at 940 °C for 24 h. One of groups of samples was quenched in oxygen gas such that its temperature dropped suddenly from 940 °C to room temperature in about 45 seconds. The other group were left in the oven to cool down slowly at a rate about 1-2 °C/min in flowing air. The two groups of samples were examined by XRD using a fully automatic Philips diffractometer (PW1170) with Co target and Ni filter. Temperature- resistivity measurements were carried out by a standard four-probe method using a Keithley nanovoltmeter model 181 and a d.c. Keithley current source model 244 with a constant current value 10 mA. A PT-100 thermocouple was used for temperature measurements connected with a digital micro-voltmeter "Dynascan 2831". Morphology of the fracture surface grains of the samples were investigated using scanning electron microscope (JEOL-JSM-5410 Tokyo, Japan) equipped with EDX unit (ANCK, Oxford, England). The bulk densities of the samples were measured using Archimedes method. While the degedation of sample in moisture were evaluated via immersing the sample in water at constant temperature (25 °C) for about one month. Then the morphology of the sample surface was studied by an optical simple microscope every 12 hours. Once a change in the color of the sample or the appearance of white spots on the sample surface give indication for the starting of sample degradation and the time for the occurrence of such phenomenon was recorded.

3. RESULTS AND DISCUSSION

3.1. X-ray Diffraction and electron microscopy

3.1.1. Samples prepared by slow cooling procedure:

The system $[YBa_2Cu_3O_{7-y}]_{1-x}(In_2O_3)_x$ where x = 0.0, 0.3, 0.6, 0.9, 1.5, 2, 4, 5 and 7 % was prepared by slow cooling method and different phase formation during sintering were characterized using XRD analysis. The XRD patterns of all samples were shown in Fig. 1. It is clear that the pure sample (x = 0.0 % wt) exhibited a single 123 orthorhombic phase $YBa_2Cu_3O_{7-y}$. The main characteristic diffraction peaks for this phase was recognized at 2θ = 26.59, 32.53, 37.99, 38.35, 45.06, 47.30, 54.76, 55.76, 68.86, and 69.60 [15]. On the other hand, the XRD patterns of samples produced with different In_2O_3 concentations revealed that thee is a partial decomposition of the 123 superconducting phase with the formation of one or more new secondary phases. This is clearly noticed from the slight decrease in the diffraction peak intensities of the 123 superconducting phase with the appearance of a new diffraction peaks. The new diffractions lines appear at the diffraction angles 2θ = 32.50, 34.56, 35.12, 35.60, 48.76, and 49.60. The diffraction peaks appear at the diffraction angels 2θ =34.56, 35.12 and 48.76 was found to be corresponding to the existence of $Ba_3In_2O_6$, (Tri-barium di-indium oxide) with tetragonal unit cell and lattice parameters a = b = 4.192 A, c = 21.689 A [16]. Whereas, the diffraction peaks appear at 2θ = 32.50, 34.56, 35.60, and 49.60 was found to be corresponding to the formation of Y_2BaCuO_5 (Barium Copper Yttrium Oxide), which is a semi-conducting compound with an orthorhombic unit cell and lattice parameters a = 7.131 A, b = 5.659 A and c = 12.18 A [15].

It is worth to mention that the peak intensities of the new phases $Ba_3In_2O_6$ and Y_2BaCuO_5 were found to increase with increasing In_2O_3 content, especially for the sample of In_2O_3 content > 1.5 wt %. Meanwhile, for samples produced with low In_2O_3 concentrations ($x < 1.5$) the peaks of the two phases were too weak to be recognized in the diffraction patterns.

Although the value of the ionic radius of In atom (0.80 A) is comparable to that of Cu^{+1} (0.77 A) and Cu^{+2} (0.73 A) we couldn't find any evidence in the XRD analysis that there is a partial replacement of any of the 123 unit cell atoms by In atom. The variation of the unit cell lattice parameters (a, b) and (c) for all samples with different In_2O_3 content was shown in Fig. 2a and 2b, respectively. It could be noticed that there were no significant changes in the unit cell lattice parameters of the superconducting phase $YBa_2Cu_3O_{7-y}$ due to the increase in the In_2O_3 content. This confirms that the addition of In_2O_3 make only partial decomposition for the 123 superconductive phase without replacement of any of 123 unite cell atoms by In atom.

However, Meissner Magnetic Levitation Test was found to be positive for all the liquid nitrogen cooled samples. The magnet was repelled and suspended in air over the sample surface. This was observed for all samples including the sample with the highest In_2O_3 content (7 % wt). This is rather confirmation that the secondary phases formed with the addition of different In_2O_3 concentration are just coexist with the parent 123 orthorhombic phase without affecting its orthorhombicity or oxygen stoichiometry. This results is in a good agreement with that deduced from the XRD study.

3.1.2. Scanning Electron Microscope (SEM) analysis

Scanning electron micrographs for the samples of x = 0.0, 0.3, 2, and 4 % were shown in Fig. 3. The pure sample (x = 0.0 %), Fig. 3a, shows that the samples contains a significant amount of pores, with randomly oriented grains of prism-like structure and smooth surfaces, which represent the characteristic features of the 123 superconducting phase [15,17]. For all other samples with different In_2O_3 content ($x > 0.0$ %) the samples surface was found to contain a larger (roundish prism-like) grains beside a smaller (white) grains that were expected to correspond to the formed secondary phases. The energy dispersive X-ray analysis (EDX) (Fig. 4.) of that rough small grains formed between the main grains (in the sample x = 0.9 wt%) indicates that this phase composed of different elements of Y, In, Ba, and Cu which confirms the formation of secondary phases as deduced from the XRD analysis. The volume of these rough white regions was found to increase as the In_2O_3 content in the sample increased. This observation is in a good agreement with the XRD results. Moreover, it could be noticed that the grain size of the 123 main phase didn't show any significant change with increasing In_2O_3 content.

3.2. Samples prepared by quenching procedure:

Samples of the system $[YBa_2Cu_3O_{7-y}]_{1-x}(In_2O_3)_x$ with x = 0.0, 0.9, 2.0, and 4.0 wt% were quenched in oxygen gas such that their temperature dropped from sintering temperature at 940 °C to room temperature in less than one minute. All quenched samples were tested using the Meissner magnetic levitation test. It was found that the test was positive i.e. the magnet was strongly repelled and suspended in air over all samples with In_2O_3 content ($x > 0.0$ wt%). This confirms the formation of the orthorhombic superconducting phase during the very short time taken by the quenching process. On the other hand and as expected, the pure sample (x = 0.0 wt%) didn't positively respond to the test and the magnet didn't fly over it. This negative result confirmed that the 123 orthorhombic phase didn't form in this sample during quenching.

Figure. 5 shows X-ray powder diffraction patterns for the four quenched samples. It is revealed that the pure sample (x = 0.0 wt%) sample showed a tetragonal 123 (non-superconducting) phase. This is clear in the peak changes at the angles 2θ = 37.99, 38.36, 54.78, 55.76, 68.86 and 69.57 [2,5-7,18]. All the other samples (x > 0.0 wt%) showed all the characteristic peaks of the orthorhombic $YBa_2Cu_3O_{7-y}$ superconducting phase with the appearance of new phases corresponding to the presence of both $Ba_3In_2O_6$ and Y_2BaCuO_5 as found in the slowly cooled samples.

However, in order to investigate the reason of such behavior for In_2O_3 content samples, the phase composition differences between the surface and interior of the samples produced with In_2O_3 addition of 2 and 4 wt% were investigated with XRD analysis. To identify the phase composition of the sample interior, both samples (with 2 and 4 wt% In_2O_3 content) was polished to the depth of about 2.5 mm and then the polished surface is subjected to the XRD analysis. Figures 6A and 6B show the XRD patterns for both 2 and 4 wt% In_2O_3 content samples, respectively. The patterns for both samples revealed that the sample surface had a clear orthorhombic123 superconductive phase, while the sample interior is composed of a clear tetragonal 123 phase.

Since the rates of oxygen absorption in pure 123 phase are known to be low at temperatures below 700 °C, slow cooling in flowing oxygen gas is an essential condition for obtaining the orthorhombic (superconducting) phase $YBa_2Cu_3O_{7-y}$ [2-4, 19-23]. The phase transformation of the material $YBa_2Cu_3O_{7-y}$ from tetragonal to orthorhombic takes place slowly below 700 °C by absorbing oxygen during slow cooling (1-2 °C/min.) or during several hours of annealing at a fixed temperature in the range (500- 600 °C). It is an established experimental fact that a sample which is quenched in air or in oxygen from a temperature above 700 °C to room temperature is always found to have the tetragonal (semi-conducting) phase and never showed the 123 orthorhombic (superconducting) phase [2,5-7]. However, the obtained results is disagreed with this fact, the sample surface was found to be composed of 123 orthorhombic (superconducting) phase. The appearance of this phase may be attributed to the presence of secondary phases which formed near the grains of 123 phase. There is a common belief that the presence of secondary phases among the 123 grains are responsible for enhancing the oxygen absorption during the quenching process. So, during quenching, the secondary phase enables the ample to absorb oxygen with a high rate (sufficient for the occurrence of phase transformation) in just one minute [8-14]. In contrary, the sample interior was found to be composed of 123 tetragonal (semi-conducting) phase. This indicates that the enhancing in oxygen absorption during sample quenching is only working in the outer layers of the sample that are more nearer to the atmospheric oxygen. On the other hand, this phase difference between the sample surface and its interior also proved that the secondary phase didn't play the role of the oxidizing agent for the tetragonal phase as some authors suggested [5,10,11], i.e. it wasn't the source of oxygen that was gained by the tetragonal phase during quenching process since the samples interior didn't get such oxygen and stayed tetragonal. To rather confirm this explanation, the same samples was allowed to be slowly cooled in the furnace and again the phase composition of sample surface and its interior was identified (Fig 7A and 7B). The XRD patterns of both samples (x = 2 and 4 wt%), show a single orthorhombic phase in both their sample surface and its interior.

Accordingly, this method of rapid quenching of sample in oxygen atmosphere could be recommended in preparation a small thickness or thin film of 123 superconductive material. This suggestion needs to be experimentally verified to make sure that the resulting samples will not suffer from micro-(or macro) cracks during thermal quenching process.

3.3. Electrical Transport

The temperature dependence of the normalized resistivity (ρ/ρ_o, %) for the system $[YBa_2Cu_3O_{7-y}]_{1-x}(In_2O_3)_x$ was studied. The results were shown in Fig. 8, for the samples x = 0.0, 0.3, 0.6, 0.9, 1.5, 2, 4 and 7 % by weight. The pure sample (x = 0.0 wt%) shows the expected metallic behavior in the normal state followed by a narrow transition to the superconducting state with Tc(onset ~ 100 K) and Tc (offset) = 91 K. Also samples with various In_2O_3 content show a metallic behavior in the normal state and the Tc(onset) values for all of them were (~ 100 K). The measured Tc(offset) values for the samples with In_2O_3 content < 1.5 % didn't show a significant change than that of the pure sample (i.e. ~ 90- 91 K). Meanwhile for samples with higher In_2O_3 content ($4 \geq x \geq 1.5$ % wt), a slight decrease in Tc-values could be observed. On the other hand the significant decrease in the value of Tc(offset) is only observed for the sample with the highest In_2O_3 content (x = 7 %) as shown in Fig. 9. As previously described, these samples (x > 0.0 wt%) composed of three phases. The first is the parent 123 orthorhombic phase whose grains constitute the majority of the sample structure and its behavior in normal state is metallic [15,24]. The other two phases are the secondary phase $Ba_3In_2O_6$ whose electrical behavior in normal state is unknown and the green phase Y_2BaCuO_5 whose behavior is semiconducting. The general metallic behavior of the system samples in normal state indicate that the electrical behavior of the new phase $Ba_3In_2O_6$ is either metallic or insulator (i.e. not semiconducting). Also the volume of the green phase Y_2BaCuO_5 in the samples is too small to observe its semi-conducting behavior in normal state.

It is well known that when the temperature decreased until the critical temperature Tc is reached the phase $YBa_2Cu_3O_{7-y}$ in the main grains becomes in the superconducting state. In the pure sample (x = 0.0 %), cooper pairs tunnel through the weak links among the grains successfully and the d.c. resistance measured across the sample drops to zero. The measured decrease in Tc- values in the samples with In_2O_3 content ≥ 1.5 wt% can be attributed to the combined effect of the two secondary phases. The presence of these phases in the inter-granular spacing between the grains plays the role of electrical barriers that must be tunneled by electron pairs in super-conduction state [25-27]. The critical current Ic of the whole structure is determined by the josephson tunneling currents flowing in the barriers between the grains [28]. This critical current decreases with increasing the thickness of potential tunneling barriers between the grains and at a certain limit it vanishes. In this case if the temperature of the sample is decreased lower than Tc the electron pairs succeed in tunneling through these barriers and superconduction current is restored again [24,29,30]. Indeed, the observed decrease in Tc values was not too large which could attributed to the fact that the volume of the main 123 orthorhombic grains didn't show a large changes with increasing In_2O_3 content (as shown in Fig. 3). Accordingly the average barrier thickness is reduced and allowed the samples to have approximately the same Tc values. The large decrease in the Tc values was only observed for sample with highest In_2O_3 content could attributed to the high increase in the volume of the secondary phase present neighboring to the superconductive phase.

However, a gradual increase in the values of the transition width ΔT for samples $x \geq 1.5$ is clearly noticeable (Fig. 10). This increase in ΔT can not be considered as an indicator for the presence of superconducting multi-phases in the sample. It can be just attributed to the decrease in Tc-values which resulted from the tunneling problems among the grains.

The temperature dependence of the normalized resistivity (ρ/ρ_o %) for the four quenched samples x = 0.0, 0.9, 1.5 and 2 wt% is shown in Fig. 11. The pure sample (x = 0.0 wt%) not showing metallic behavior as a result of the absence of the 123 orthorhombic (superconductive) phase (Fig. 5), while other samples with various In_2O_3 content (x = > 0.0 wt%) show a metallic behavior in the normal state and then a sudden transition to the superconducting state with Tc (onset) nearly equal to its corresponding values for the slowly cooled samples. The Tc(offset) values didn't show any significant changes and its values were around those measured in the slow cooled samples.

3.4. Bulk Density Measurements
The variation in the bulk density of sample produced with various In_2O_3 content for both slowly cooled and quenched samples is shown in Fig. 12. For the slowly cooled samples, it was found that increasing In_2O_3 content leads to a significant increase in the bulk densities for the samples $x \leq 4$ % wt. This was attributed to the fact that with increasing In_2O_3 content the volume of the secondary phases which formed among parent phase grains is gradually increase without changing in the grain size of the parent 123 phase. This leads to a great reduction in the number of pores produced in the final structure and consequently increasing the sample bulk density. The highest bulk density (5.3428 g/cm^3) was observed for sample with In_2O_3 content of 4 wt%. While with further increasing in the In_2O_3 content beyond this percentage (at x = 5 wt%) a slight decrease in the bulk density was noticed. This drop in the bulk density indicating that there is some sort of reduction in the grain size of the parent phase were occurred with the further increase in the In_2O_3 content beyond 4 wt%. The same behavior was observed for the quenched samples unless it is gradually increased with increasing In_2O_3 content even for the highest In_2O_3 content (x = 5 wt%). On the other hand the bulk densities for the quenched samples was found to be generally lower than those for the slowly cooled samples. This could attributed to the formation of some cracks during thermal quenching.

3.5. Sample degradation in water
Both slowly cooled and quenched In_2O_3 content samples were immersed in water at a fixed temperature 25 °C for one month. The samples surface was thoroughly checked every 12 hours. The surface of pure sample (x = 0.0 wt%) had white spots and a temporary dark green color after 2.5 days (Fig. 13). The green color indicated the formation of the green phase, Y_2BaCuO_5, and the white spots referred to the formation of $BaCO_3$ [31]. The water surface had a white color thin film which is thought to be formed by $BaCO_3$. However, the In_2O_3 content samples was found to need longer times to have such changes. The longest period was recorded for sample with In_2O_3 content = 2 wt%. This sample kept its pure black color (without white spots) for 15 days. In the day 16th it started to have the white spots and the water surface formed a white thin film. Samples with higher In_2O_3 content (x > 2 %) kept their dark black surfaces for shorter periods.

Along a whole month immersed in water, all the samples responded positively to Meissner magnetic levitation test indicating the presence of the superconducting phase. None of the samples was broken in water except for the pure one (x = 0.0 wt%).

The quenched samples were also immersed in water and their surface morphology was checked every 12 hours. The pure sample (x = 0.0 %) had white spots just after 1 day. While, samples with In_2O_3 content started to lose their black color in periods relatively shorter than those observed for their corresponding slowly cooled samples. This can be attributed to the formation of micro-cracks during quenching procedure [32].

CONCLUSION
The addition of various percentage of In_2O_3 to the 123 orthorhombic (superconductive) material leads to formation of secondary phases of $Ba_3In_2O_6$ and Y_2BaCuO_5 for samples slowly cooled as well as for the quenched samples. These secondary phases are co-exist with the parent orthorhombic phase without affecting its orthohombicity or oxygen stoicheometry. The presence of these phases enhance the rate of oxygen absorption (for quenched samples) especially for the sample surface that are nearer to the oxygen atmosphere. This causes the formation of 123 orthorhombic (superconducting) phase on

the surface of the In_2O_3 content samples. Also adding small amounts of In_2O_3 to the pure 123 phase increases its bulk density and improves its resistance to corrosion in humidity. It was found that the optimum value of the In_2O_3 ratio is (x = 2 % by weight). This ratio causes an increase of the sample bulk density by 15.7 % and also improves its resistivity to corrosion in water 6 times.

REFERENCES

1. K.V. Paulose, J. Koshy, and A.D. Damodaran, " Observation of Superconductivity in Nb_2O_5 Doped $YBa_2Cu_3O_{7-\delta}$ Compound by Rapid Quenching," Jpn. J. Appl. Phys. **30**, L458-L460 (1991).
2. W.E. Farneth, R.K. Bordia, E.M. McCarron, M.K. Crawford and R.B. Flippen, "Influence of oxygen stoichiometry on the structure and superconducting transition temperature of $YBa_2Cu_3O_x$," **66 [9]**, 953-959 (1988).
3. J. D. Jorgensen, M. A. Beno, D. G. Hinks, L. Soderholm, K. J. Volin, R. L. Hitterman, J. D. Grace*, and Ivan K. Schuller, "Oxygen ordering and the orthorhombic-to-tetragonal phase transition in $YBa_2Cu_3O_{7-x}$," Phys. Rev. B 36, 3608–3616 (1987)
4. W. Wong-Ng, L. P. Cook, C. K. Chiang, L. J. Swartzendruber, L. H. Bennett, J. Blendell, D. Minor, "Structural phase transition study of $Ba_2YCu_3O_{6+x}$ in air," J. Mat. Res. 3, 832-839 (1988).
5. H. Ihara, H. Oyanagi, R. Sugise, E. Ohno, T. Matsubara, S. Ohashi, N. Terada, M. Jo, M. Hirabayashi, K. Murata, A. Negishi, Y. Kimura, E. Akiba, H. Hayakawa, S. Shin, "Strong correlation between T_c and lattice parameters in $Ba_2YCu_3O_y$," Physica C., 153-155 [2], 948-949 (1988).
6. E.T. Muromachi, Y. Uchida, M. Ishii, T. Tanaka and K. Kato, "High T_c Superconductor $YBa_2Cu_3O_y$–Oxygen Content vs T_c Relation," Jpn. J. Appl. Phys. 26, L1156-L1158 (1987).
7. M. Tokumoto, H. Ihara, T. Matsubara, M. Hirabayashi, N. Terada, H. O., K. Murata and Y. Kimura, "Evidence of Critical Oxygen Concentration at y=6.7~6.8 for 90 K Superconductivity in $Ba_2YCu_3O_y$, " Jpn. J. Appl. Phys. 26, L1565-L1568 (1987).
8. K.V. Paulose, J. Koshy and A.D. Damodaran, "Observation of Superconductivity in Nb_2O_5 Doped $YBa_2Cu_3O_{7-\delta}$ Compound by Rapid Quenching," Jpn. J. Appl. Phys., 30, L458- L460 (1991).
9. S.M. Rao, J.K. Srivastava, M.J. Wang, S.R. Sheen, M.K. Wu and Y.F. Chen, "Oxygen stabilization in $YBa_2Cu_3O_{6+\delta}$:Ta_2O_5 sintered in air," Supercond. Sci. Technol. 13, 1264–1269 (2000).
10. Y. Feng, L. Zhou, S. Du and Y. Zhang, "Study of quenched $YBa_2Cu_3O_{7-y}$-WO_3 superconductors," J. Phys. Condens. Matter., 6 [45] 9755-9758 (1994).
11. G. Nieva, B.W. Lee, J. Guimpel, H. Iwasaki, M.B. Maple and I.K. Schuller, "Changes in the Pr-induced T_C depression of 123 compounds by chemical pressure," Physica C., 185-189, 561-562, (1991).
12. K.V. Paulose, J. Koshy and A.D. Damodaran, "Superconductivity in $YBa_2Cu_3O_{7-\delta}$-ZrO_2 systems," Supercond. Sci. Technol., 4, 98-101 (1991).
13. J. Koshy, J.K. Thomas, J. Kurian, Y.P. Yadava and A.D. Damodaran, "The structural and superconducting properties of the YBa,Cu,$O_{7-\delta}$-HfO_2, systems", J. Appl. Phys. 73, 3402-3407 (1993).
14. J.K. Thomas, J. Koshy, J. Kurian, Y.P. Yadava and A.D. Damodaran, "Electrical transport and superconductivity in $YBa_2Cu_3O_{7-\delta}$ -$YBa_2HfO_{5.5}$ percolation system," J. Appl. Phys. 76, 2376-2380 (1994).

15. C.P. Poole, T. Datta and H.A. Farach, "Copper Oxide Superconductors," Wiley & Sons Inc. USA, (1988).
16. P.W. Anderson, "Theory of Flux Creep in Hard Superconductors," Phys. Rev. Lett. 9, 309-311 (1962).
17. Rizzo Assuncao, F. C. et al, "Effect of Niobium substitution in superconducting copper-based oxides,". Symposium on High-Temperature Superconductors. Structure and Microstructure, Bad Nauheim, (FRG) April 21-22, (1988).
18. D.M.R. Lo Cascio, M.T. Van Wees and H. Bakkar, "Influence of quenching speed on the superconducting transition temperature of $YBa_2Cu_3O_{7-\delta}$ annealed in air at 650–1200 K," J. Appl. Phys., 71, 1885-1888 (1992).
19. R.J. Cava, R.B. Van Dover, B. Batlogg and E.A. Rietman, "Bulk superconductivity at 36 K in $La_{1.8}Sr_{0.2}CuO_4$," Phys. Rev. Lett. 58, 408-410 (1987).
20. P.K. Gallagher, H.M. O'Bryan, S.A. Sunshine and D.W. Murphy, "Oxygen stoichiometry in $Ba_2YCu_3O_x$," Mater. Res. Bull. 22, 995-1006 (1987).
21. K. Kishio, J. Shimoyama, T. Hasegawa, K. Kitazawa and K. Fueki, "Determination of Oxygen Nonstoichiometry in a High-T_c Superconductor $Ba_2YCu_3O_{7-\delta}$," Jpn. J. Appl. Phys. 26, L1228-L1230 (1987).
22. J. D. Jorgensen, B. W. Veal, A. P. Paulikas, L. J. Nowicki, G. W. Crabtree, H. Claus, and W. K. Kwok, "Structural properties of oxygen-deficient $YBa_2Cu_3O_{7-\delta}$," Phys. Rev. B., 41, 1863-1877 (1990).
23. S. Sugai, "Effects of oxygen deficiency on the infrared spectra in $YBa_2Cu_3O_{7-\delta}$," Phys. Rev. B., 36, 7133-7136 (1987).
24. F. Yong, Z. Lian, Shejun, Du, Jinrong, Wang and Z. Yuheng, Physica C 235-240 405-406 (1994).
25. C.P. Poole, H.A. Farach and R.J. Creswick, "Superconductivity" Academic Press, USA (1995).
26. M. Cyrot and D. Pavuna, "Introduction to Superconductivity and High-Tc Materials," World Scientific Publishing Co. Pte. Ltd. Singapore (1992).
27. K.V. Kumar and T. Sreekanth, "Solid State Physics," S. Chand & Company Ltd, New Delhi (2005).
28. B. Raveau, C. Michel, M. Hervieu and D. Groult, " Crystal Chemistry of High-Tc Superconducting Copper Oxides," (Springer Series in Materials Science, Vol. 15), p 331, Germany, Springer, (1991).
29. S. Sergeenkov and M. Ausloos, "Thermoelectric power of inhomogeneous superconductors: Effects of field-induced intragrain granularity," Phys. Rev. B. 47, 14476-14480 (1993).
30. M.D. Fiske, "Temperature and Magnetic Field Dependences of the Josephson Tunneling Current," Rev. Mod. Phys., 36, 221-222 (1964).
31. J.P. Zhou, J.T. McDevitt "Corrosion Reactions of $YBa_2Cu_3O_{7-X}$ and Tl2ba2ca2cu3o10+X Superconductor Phases in Aqueous Environments," Chem. Mater., 4, 953-959 (1992).
32. T.M. Shaw, S.L. Shinde, D. Dimes, R.F. Cook, P.R. Duncombe, and C. Kroll, "The effect of grain size on microstructure and stress relaxation on polycrystalline$YBa_2Cu_3O_{7-y}$," J. Mater. Res., 4, 248-256 (1989).

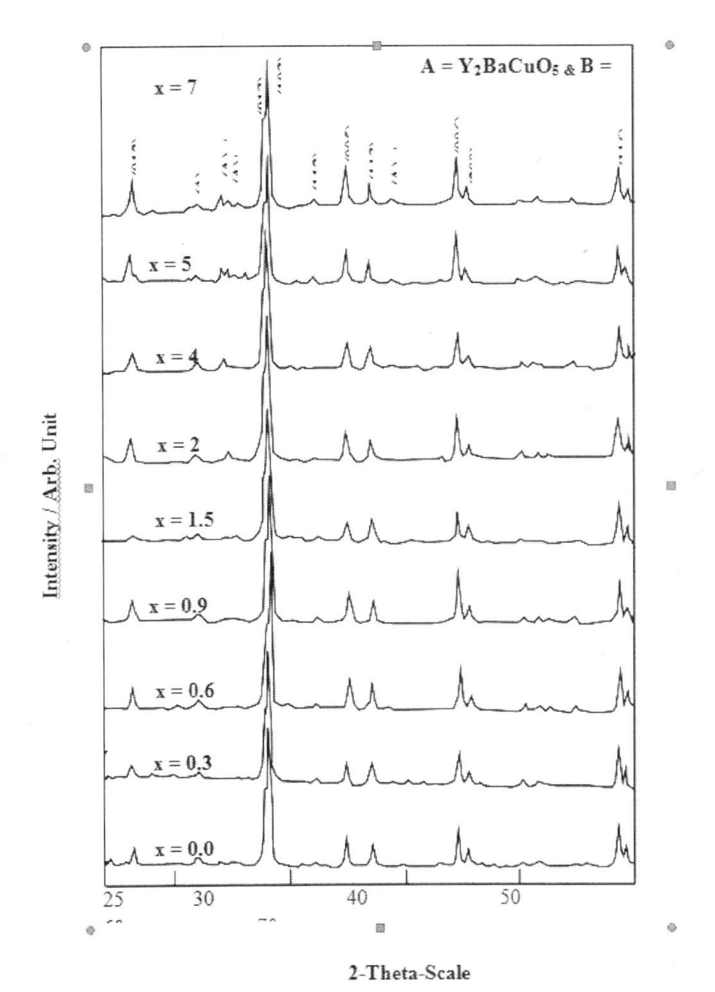

Fig. 1. XRD patterns for $[YBa_2Cu_3O_{7-y}]_{1-x}(In_2O_3)_x$ system.

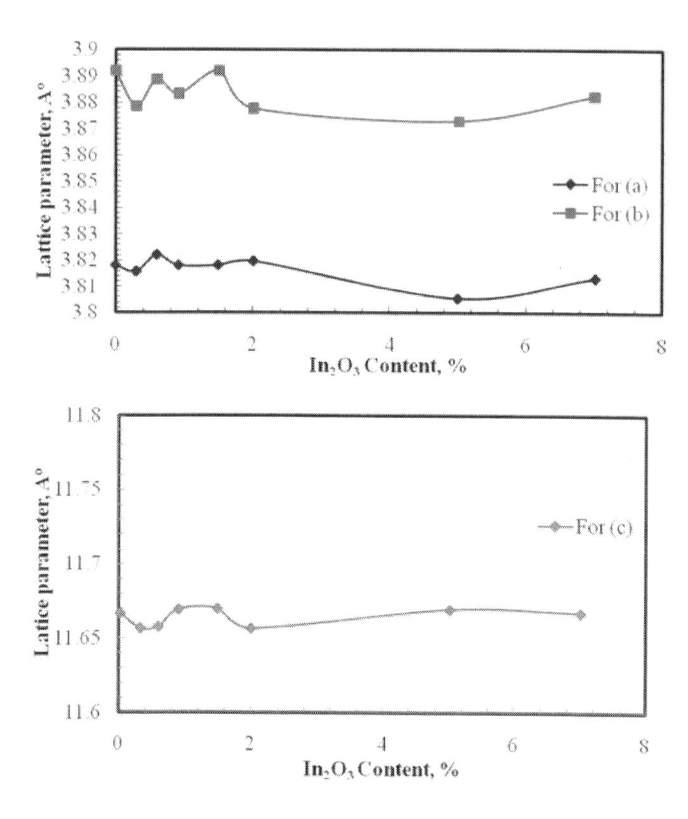

Fig. 2. Lattice parameters for a, b and c of the In_2O_3 doped samples

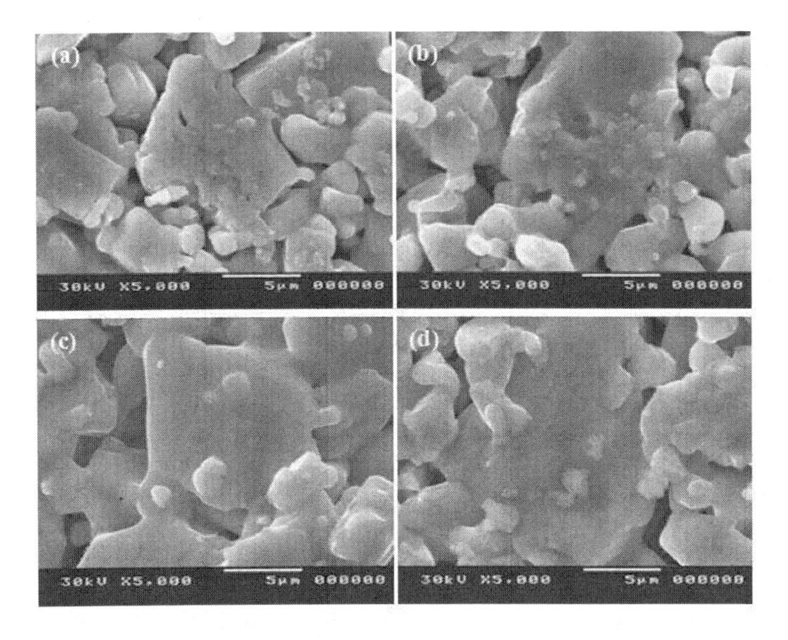

Fig. 3. Scanning electron micrographs for the doped samples

(a) for x = 0.0%, (b) for x = 0.9 %, (c) for x = 2% and (d) for x = 4%

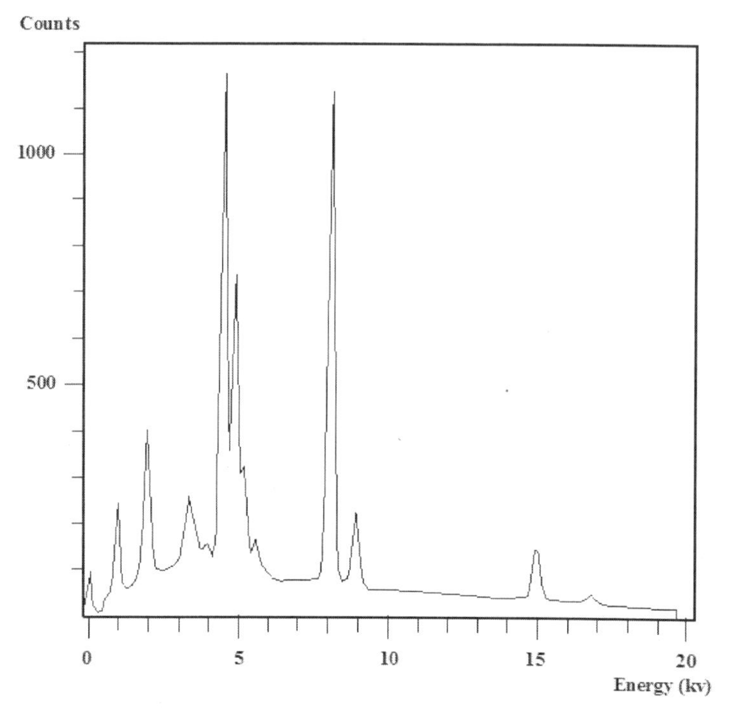

Fig. 4. Energy dispersive x-ray analysis (EDX) of the rough regions in the In_2O_3 doped sample (x = 0.9 %).

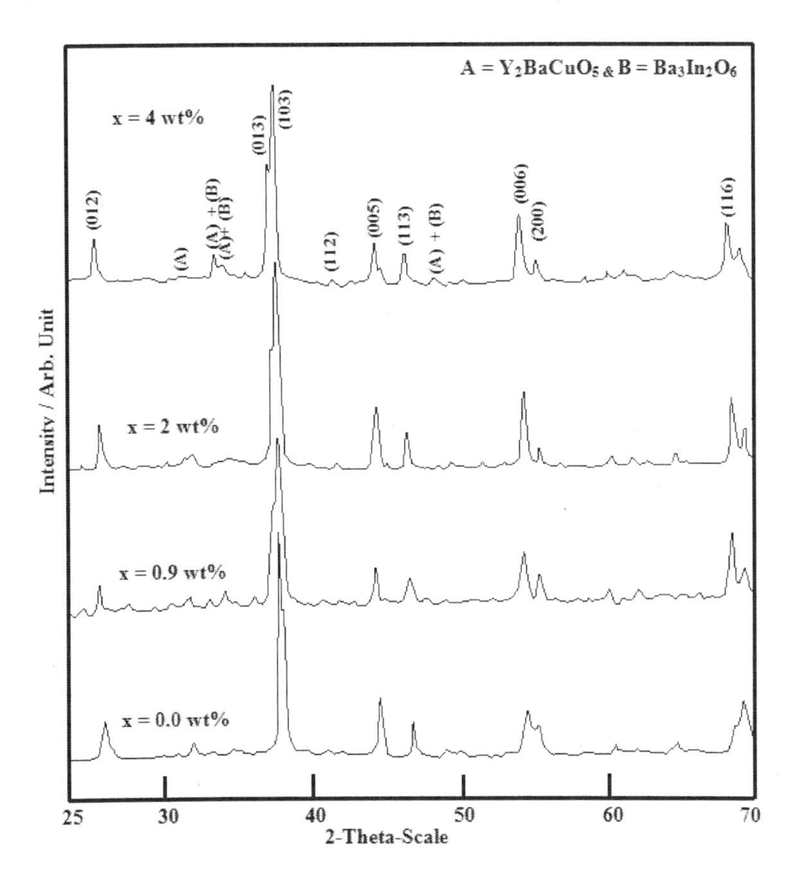

Fig. 5. XRD pattern for the $[YBa_2Cu_3O_{7-y}]_{1-x}(In_2O_3)_x$ system quenched from 940 °C to room temperature in oxygen gas: (a) 0.0 %, (b) 0.9 %, (c) 2 % and (d) 4 %.

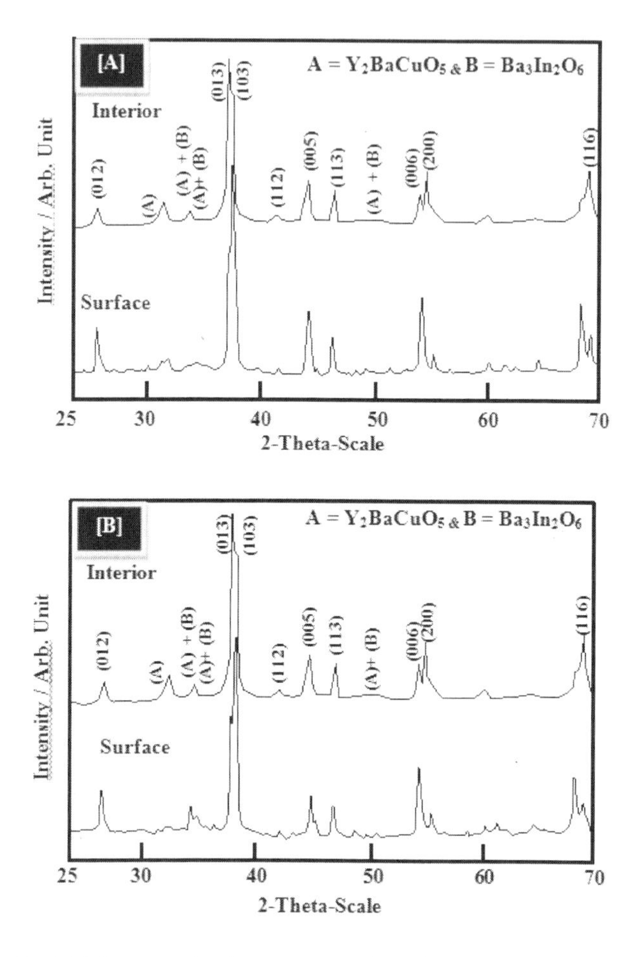

Fig. 6. XRD patterns for the In_2O_3 - doped samples quenched in oxygen gas.

A is sample doped with 2% In_2O_3, B is sample doped with 4% In_2O_3

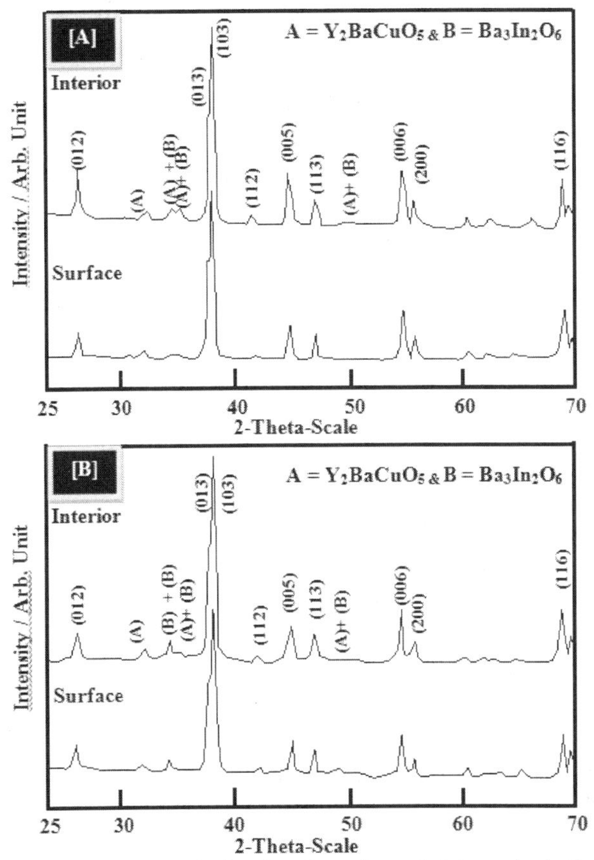

Fig. 7. XRD patterns for the In_2O_3 - doped samples cooled down slowly from 940 C to room temperature at a cooling rate 2 C/min.

A is sample doped with 2% In_2O_3, B is sample doped with 4% In_2O_3

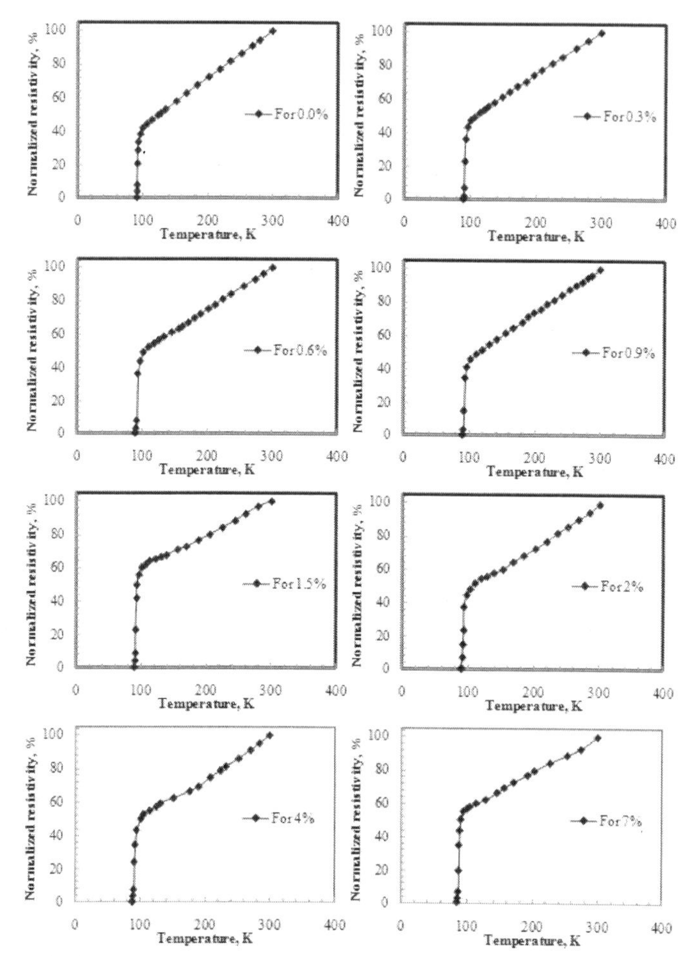

Fig. 8. Temperature dependence of normalized resistivity (ρ/ρ_o) % for the system $[YBa_2Cu_3O_{7-y}]_{1-x}(In_2O_3)_x$ for different x values (a) 0 %, (b) 0.3 %, (c) 0.6 %, (d) 0.9 %, (e) 1.5 %, (f) 2 %, (g) 4 % and (h) 7 %.

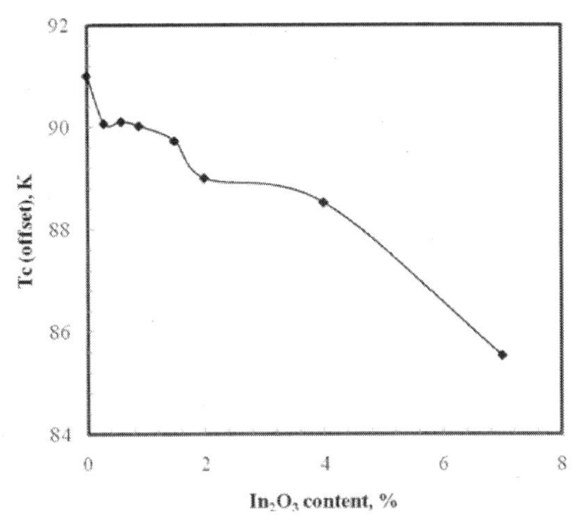

Fig. 9. Variation of Tc-values for the In_2O_3-doped system with In_2O_3 content.

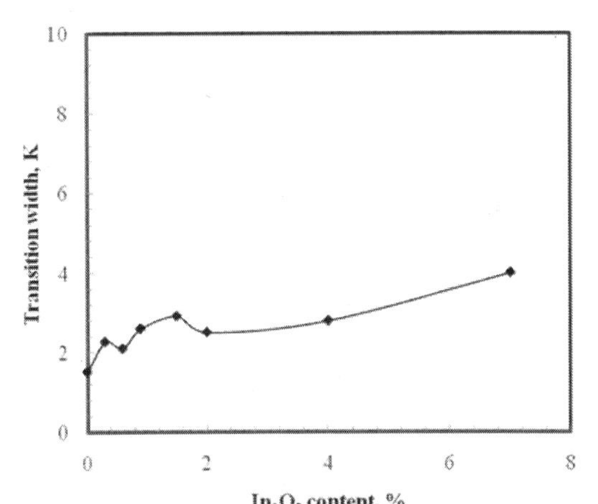

Fig. 10. Variation of transition width ΔT with In_2O_3 content

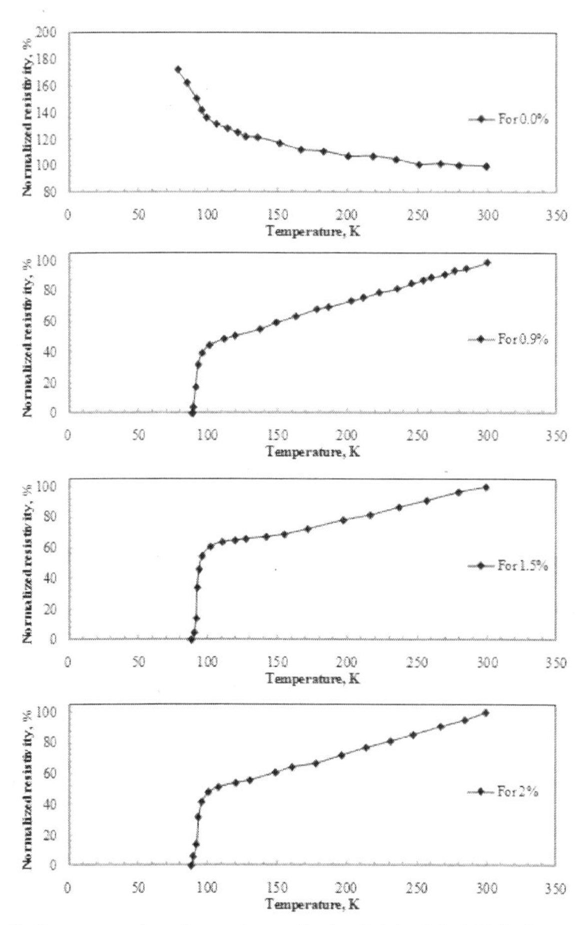

Fig. 11. Temperature dependence of normalized resistivity (ρ/ρ_o) % for the system $[YBa_2Cu_3O_{7-y}]_{1-x}(In_2O_3)_x$ quenched from 940 C to room temperature in oxygen gas: (a) 0.0 %, (b) 0.9 %, (c) 1.5 %, and (d) 2 %.

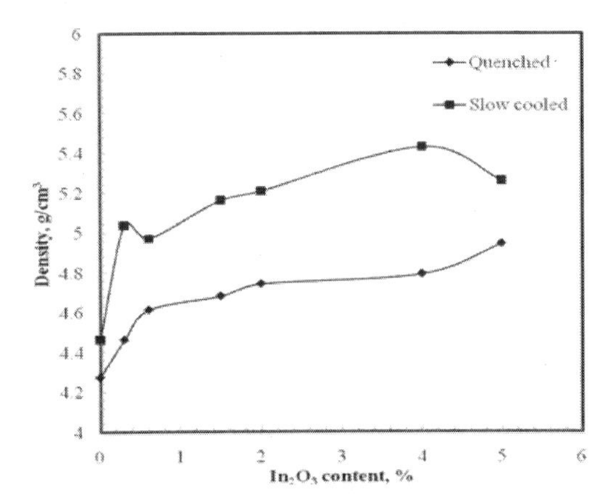

Fig. 12. Bulk density for quenched and slowly cooled samples of the system $[YBa_2Cu_3O_{7-y}]_{1-x}(In_2O_3)_x$.

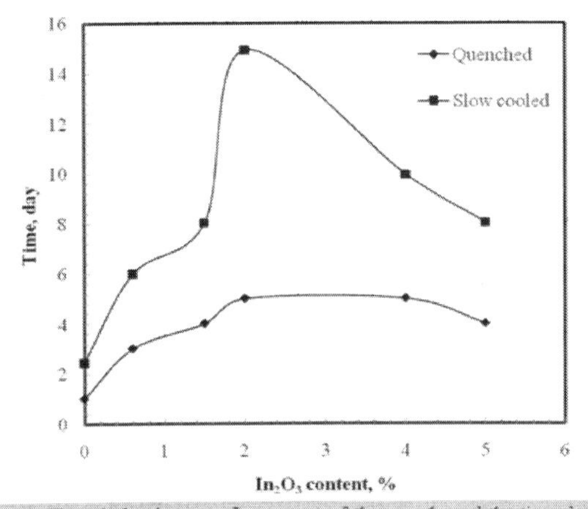

Fig. 13. The relation between In-content of the sample and the time during which it could stay immersed in water without the appearance of corrosion symptoms.

MORPHOLOGIES AND ELECTROCHEMICAL CAPACITOR BEHAVIORS OF $Co(OH)_2$/POLYANILINE COMPOSITE FILMS

Hiroshi Itahara and Tetsuro Kobayashi
Toyota Central R&D Labs., Inc
Nagakute 41-1, Aichi, Japan

ABSTRACT

We prepared the $Co(OH)_2$ or $Co(OH)_2$/polyaniline composite film by the electrolytic deposition and evaluated their specific capacitance (Cp) by the cyclic voltammetry (CV). We examined the effect of deposition conditions on morphologies of the $Co(OH)_2$ particles in the film, where $Co(NO_3)_2$ aqueous solution without or with the water-soluble conducting polyaniline-derivative was used as the raw material. From the solution without polyaniline-derivative, the scale-like α-$Co(OH)_2$ particles were obtained at 303 K while rather thicker β-$Co(OH)_2$ platelets were formed at 348 or 363 K. The α-$Co(OH)_2$ film showed Cp of 790 F/g at a scan rate of 2 mV/s, using 0.1 M KOH aqueous solution as an electrolyte. In contrast, the β-$Co(OH)_2$ film with thicker particles showed much lower Cp of 41 F/g. The difference in Cp values between α- and β-$Co(OH)_2$ films may be attributed to their crystallinity and morphologies. On the other hand, the solution with polyaniline-derivative at 303 K gave the deposit film composed of diminished sized α-$Co(OH)_2$ particles and polyaniline-derivative. The Cp of the composite film was not enhanced to the value expected for the materials with increased the electrode-electrolyte interface area. Here, the Co valency state was reported to change through insertion or extraction of proton. CV with various scan rates indicated a possible suppression of proton diffusion for the composite film because of coexisting polyaniline-derivative. These results suggest that the balance between proton diffusion and electron conduction be the key to enhancing Cp values.

INTRODUCTION

The redox capacitor, which is one of the electrochemical capacitors, is known as the promising energy storage devices and has been attracting much attention for the past decades. [1] Those use the redox reaction of the electrode active materials for the energy storage. The feature of the redox capacitors is their higher energy density than the electric double-layer capacitors, and their higher power density and longer life cycle than the secondary batteries.

In particular, the report on extremely high specific capacitance (Cp) for the precipitation prepared RuO_x particles (720 F/g) [2] stimulated the development of various transition metal oxides and hydroxides (transition metals such as Mn, Fe, Ni and Co) as electrode materials. These oxides/hydroxides have been expected as the alternative candidates for RuO_x because of their lower costs, however, their Cp values were rather low at around 200-400 F/g [1]. For these materials, the valency state of the metal element was reported to change through insertion or extraction of proton thereby exhibiting the electrical energy storage. The distinguishing difference between those transition metal oxides/hydroxides and RuO_x would be the proton diffusivity inside their particles: Proton insertion seems to be limited to the vicinity of the particles surface for those metal oxides/hydroxide whereas proton is able to be inserted into deep inside of the RuO_x particles. Among such transition metal oxides/hydroxides, α-$Co(OH)_2$ film prepared by the electrolytic deposition was recently reported to show exceptionally high Cp value of 860 F/g. [3]

For $Co(OH)_2$ films, the further development has been attempted through several approaches. The liquid crystal templated electrodeposition was reported to give the ordered mesoporous $Co(OH)_2$ films with Cp value of 1084 F/g. [4] High Cp value would be attributed to the broad surface area of $Co(OH)_2$ and the high accessibility of electrolyte species to the surface of $Co(OH)_2$. In another approach, the $Co(OH)_2$ particles with expanded interlayer distance were prepared by the electrolytic deposition using the intercalation agent into $Co(OH)_2$ layered crystal structure, where the intercalated species were insulating materials, such as sodium dodecyl sulfate. [5] The report gave the insight for the possible enhancement on accessibility of electrolyte species into $Co(OH)_2$ particles of the electrolytic deposit films, even though the electrochemical capacitor behavior was not reported.

In this study, we prepared $Co(OH)_2$ films by the electrolytic deposition under the various conditions. In particular, we attempt the novel synthesis procedure using $Co(NO_3)_2$ aqueous solution with water soluble polyaniline-derivative, the self-doped conducting polymer, as source materials. Here, we expect that the addition of polyaniline-derivative exhibit the following effects on $Co(OH)_2$ films; the existence of polyaniline-derivative could enlarge the electrode-electrolyte interface area and increase the electron conduction, thereby enhancing Cp values of the film. This report discusses the morphologies and electrochemical capacitor behaviors of the prepared films.

EXPERIMENTAL PROCEDURES

All films were prepared by electrolytic deposition under the condition listed in Table 1 using $Co(NO_3)_2$ aqueous solution without or with polyaniline-derivative (Mw. = ~10,000, self-doped conducting polymer, SO_3H-doped polyaniline shown in the inset of Fig. 1). Reagent grade $Co(NO_3)_2 \cdot 6H_2O$ (Wako Pure Chemical Industries, Ltd.) was used without further purification and dissolved in distilled water. The polyaniline-derivative dissolved aqueous solution (Aqua-PASS, polyaniline-derivative 5 wt% solution, TA Chemical Co., Ltd.) was mixed with the $Co(NO_3)_2$ aqueous solution. Transparent solution without any precipitates was obtained.

The films were electrodeposited onto Pt-sheet cathode electrode (~ 1.0 x 1.0 cm^2) at various temperatures (303, 348 or 363 K) and various voltages in a three-electrode system as shown in Fig. 1. The constant voltage (-500, -750 or -1000 mV) was applied between Pt cathode electrode and the Ag/AgCl reference electrode for 2 h. The crystalline phase and morphology of the deposited films were characterized by X-ray diffraction (XRD) and scanning electron microscopy (SEM). The particles composing the films were detached from the electrode and their morphologies were observed by transmission electron microscopy (TEM).

Table1 Conditions of electrolytic deposition

Name	Starting material*	temperature (K)	voltage (mV)
Co-1	$Co(NO_3)_2$ (aq.)	303	-750
Co+PANI	$Co(NO_3)_2$+PANI (aq.)	303	-750
Co-2	$Co(NO_3)_2$ (aq.)	348	-750
Co-3	$Co(NO_3)_2$ (aq.)	363	-750
Co-4	$Co(NO_3)_2$ (aq.)	348	-500
Co-5	$Co(NO_3)_2$ (aq.)	348	-1000

* $Co(NO_3)_2$ (aq.) : $Co(NO_3)_2$ aqueous solution [0.02 M, 50 ml]
$Co(NO_3)_2$+PANI (aq.) : $Co(NO_3)_2$ aqueous solution [0.02 M, 50 ml] + polyaniline-derivative aqueous solution [0.1 ml]

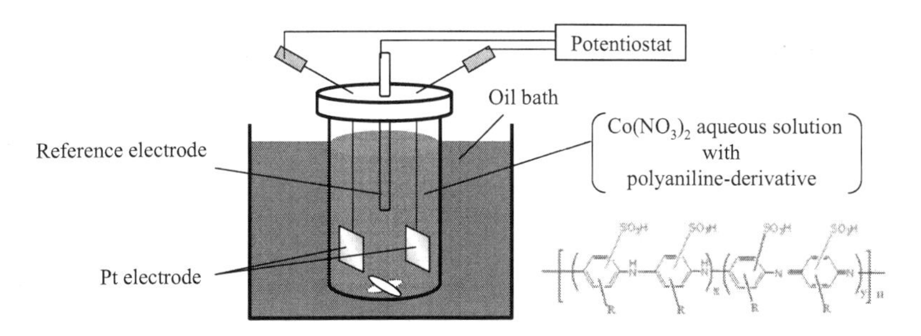

Figure 1 Experimental apparatus for electrolytic deposition

Cyclic voltammetry (CV) was conducted at various scan rates ranging from 2 to 100 mV/s, using a three-electrode electrochemical cell composed of as-prepared films on Pt sheet (working electrode), Pt sheet (counter electrode), the Hg/HgO electrode (reference electrode, which is different from that used for electrolytic deposition) and 0.1 M KOH solution as the electrolyte. Specific capacitance (Cp) of the films was calculated by the CV curve as the following equation:

$$Cp\ (F/g) = Q\ (C)\ /\ \text{potential range}\ (V)\ /\ m\ (g).$$

Here, Q and m are anodic charge and the electrode (deposit film) mass, respectively. In addition, the apparent diffusion coefficient of proton, where proton is considered to be inserted into or extracted from the deposited film during redox reactions, was estimated: We evaluated the index showing the extent of proton diffusivity using the relationship between the peak anodic current of the CV curve and the scan rates under the hypothesis of reversible proton insertion and extraction reactions.

RESULTS AND DISCUSSION

Morphology of the deposit films

Figure 2 shows the XRD patterns of the deposit films. Here, all the films were composed of small particles as described below (See, Fig. 3). At the deposition potential of -750 mV, the crystalline phase of the deposit particles was found to be affected by deposition temperature: the particles were α-$Co(OH)_2$ for the film prepared at 303 K whereas those were β-$Co(OH)_2$ at 348 and 363 K (Figs. 3 (a), (c) and (d)). Alpha-$Co(OH)_2$ formation at 303 K by the electrolytic deposition was in accordance with the previous report. [3]. In the crystal structure of $Co(OH)_2$, two polymorphs of metastable α-phase and thermodynamically stable β-phase are known. The formation of β-$Co(OH)_2$ would be preferable for higher deposition temperature. At the constant temperature (343 K), β-$Co(OH)_2$ particles were found to be formed for any deposition potentials (-500, -750, -1000 mV: XRD data are not shown). The size of the particles composing of the films were enlarged with increasing the magnitude of the deposition potential (Figs. 3(e): -500 mV, 3(c): -750 mV and 3(f): -1000 mV). The size and shape of the particles deposited at -1000 mV (Fig. 3(f)) were similar to the reported results. [5] Here, the local pH increase near the surface of the cathode electrode because of electrolytically-generated H_2 gas (OH^- ion generation in the solution) is expected to induce $Co(OH)_2$ deposition. The change in the morphology of the prepared films would be affected by the rate of the local pH increase.

As shown in Figs. 2(c)-(d), both Co-1 and Co+PANI samples show XRD patterns corresponding to α-$Co(OH)_2$ phase, where those samples were electrodeposited $Co(NO_3)_2$ aqueous solution without or with polyaniline-derivative at 303 K, respectively. Co+PANI sample was found to include polyaniline-derivative according to the SEM-EDX (energy dispersive X-ray analysis). The same d-spacing of α-$Co(OH)_2$ for those samples indicates that the polyaniline-derivative was not intercalated into the α-$Co(OH)_2$ crystal structure although both α- and β-$Co(OH)_2$ were reported to show intercalation behavior. [5-6] This result would be explained by the high molecular weight (~ 10,000) of the polyaniline-derivative. The mixture state and the amount of polyaniline-derivative in the deposit film are currently unclear, however, the water soluble polyaniline-derivative does not leak out from the films. It should be noted that the electrolytic deposition using $Co(NO_3)_2$ aqueous solution with aniline (the monomer of polyaniline) gave α-$Co(OH)_2$ particles without including polyaniline. On the other hand, the existence of the polyaniline-derivative showed the effect of diminishing the size of the deposited α-$Co(OH)_2$ particles (Figs. 3 (a)-(b) and Figs. 4 (a)-(b)). The diminished particle size is in accordance with the observed weak XRD peaks. As shown in Figs. 4(b), the Co+PANI sample are composed of particles with distorted morphologies, suggesting the depressed crystal growth of α-$Co(OH)_2$ particles by the co-deposited polyaniline-derivative. In contrast, some particles seem to be faceted for the Co-1 sample (Figs. 4(a)), even though α-$Co(OH)_2$ phase has rather amorphous-like characteristics.

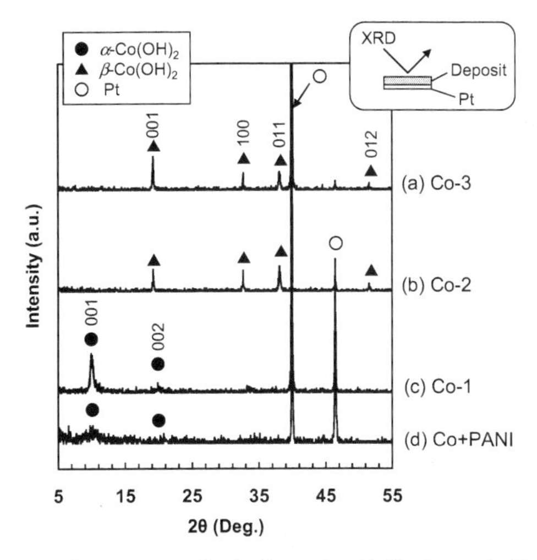

Figure 2 XRD patterns of deposit films on Pt cathode electrode: (a) Co-3 sample [deposition condition: $Co(NO_3)_2$, 363 K, -750 mV], (b) Co-2 [$Co(NO_3)_2$, 348 K, -750 mV], (c) Co-1 [$Co(NO_3)_2$, 303 K, -750 mV], (d) Co+PANI [$Co(NO_3)_2$ with polyaniline-derivative, 303 K, -750 mV]. Pt is the substrate used for the electrolytic deposition.

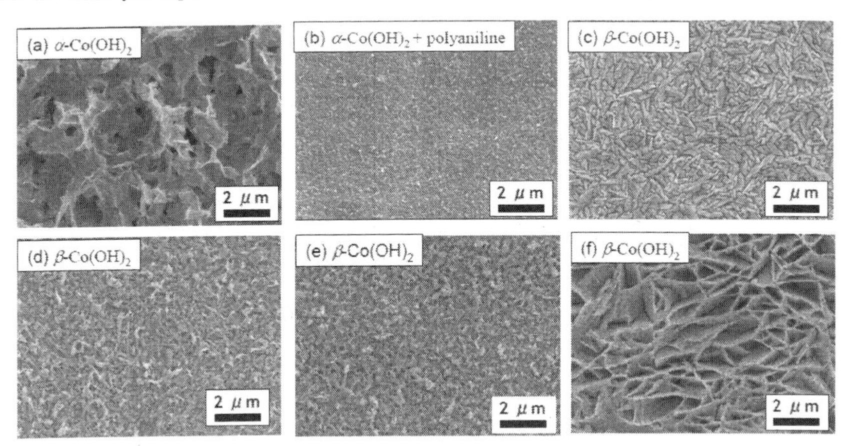

Figure 3 SEM photographs of deposit films on Pt cathode electrode: (a) Co-1 sample [deposition condition: $Co(NO_3)_2$, 303 K, -750 mV], (b) Co+PANI [$Co(NO_3)_2$ with polyaniline-derivative, 303 K, -750 mV], (c) Co-2 [$Co(NO_3)_2$, 348 K, -750 mV], (d) Co-3 [$Co(NO_3)_2$, 363 K, -750 mV], (e) Co-4 [$Co(NO_3)_2$, 348 K, -500 mV], (f) Co-5 [$Co(NO_3)_2$, 348 K, -1000 mV]

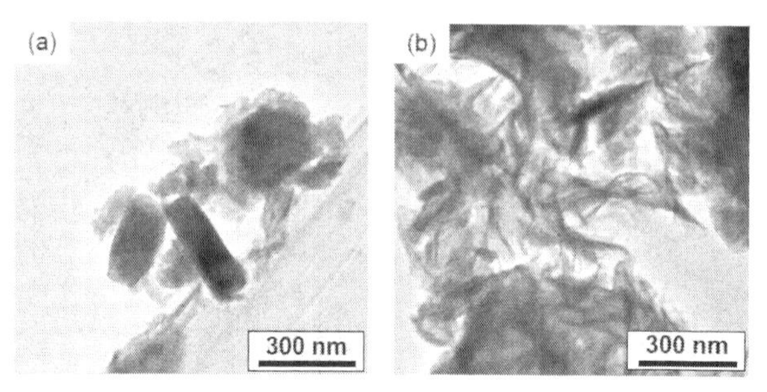

Figure 4 TEM photographs of the particles composed of the deposit films on Pt cathode electrode: (a) Co-1 sample [deposition condition: $Co(NO_3)_2$, 303 K, -750 mV], (b) Co+PANI [$Co(NO_3)_2$ with polyaniline-derivative, 303 K, -750 mV]

Electrochemical behaviors of the deposit films

The film composed of α-$Co(OH)_2$ particles (Co-1 sample) showed CV curve with strong redox peaks as shown in Fig. 5. The following redox reaction, where the Co valency state changes through insertion or extraction of proton, was suggested [3]:

$$Co(OH)_2 + OH^- \Leftrightarrow CoO(OH)_{2-} + H_2O + e^-$$

The calculated Cp was 790 F/g at a scan rate of 2 mV/s (Fig. 6). The Cp value is close to the previous report [3]. On the contrary, The film of β-$Co(OH)_2$ particles showed much lower Cp of 41 F/g (Co-5 sample: the data are not shown). As shown in Figure 3, α-$Co(OH)_2$ particles were scale-like and thus would be rather poorly crystalline than the β-$Co(OH)_2$ particles. The crystallinity may affect the Cp values. The composite film of α-$Co(OH)_2$ particles and polyaniline-derivative (Co+PANI sample) exhibited only about half value of Cp when compared to the α-$Co(OH)_2$ film (Fig. 6). The peak positions corresponding to the redox reaction were not changed between Co+PANI and Co-1 samples (Fig. 5). Thus, polyaniline-derivative itself would exhibit little redox reactions in the current system. In addition, the Cp value of Co+PANI sample was not enhanced as expected for the materials with increased electrode-electrolyte interface area: the composite film was composed of smaller sized α-$Co(OH)_2$ particles than Co-1 film. In order to consider the phenomenon behind these results, we estimated the apparent diffusion coefficient of proton inside the electrode active materials for Co-1 and Co+PANI samples as show in Figure 7. The steeper slope of the line for Co-1 sample than that of Co+PANI sample indicates that the faster proton diffusion for Co-1 sample than that for Co+PANI sample. For Co+PANI sample, the coexistence of polyaniline-derivative diminished the size of α-$Co(OH)_2$ particles thereby increasing the electrode-electrolyte interface area. Although the amount of polyaniline-derivative in the composite film was not determined, the existence of polyaniline-derivative may reduce the proton concentration diffusing inside the α-$Co(OH)_2$ particles. In addition, the depressed proton diffusion may be related also to the insufficient electrical conductivity for Co+PANI film: the electrical conductivity for the polyaniline-derivative would be decreased in reducing ambient (0.1 M KOH electrolyte), even for such a self-doped conducting polymer. The Cp values would be further enhanced by the best balancing between proton diffusion and electron conduction.

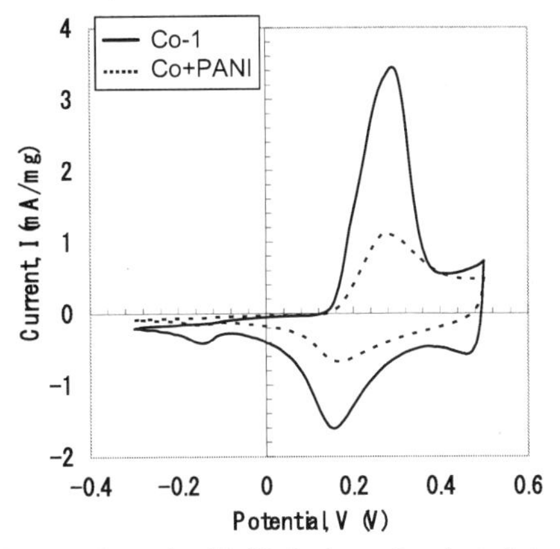

Figure 5 CV curves (reference electrode: Hg/HgO electrode, electrolyte: 0.1 M KOH) for deposit films: Co-1 sample [deposition condition: $Co(NO_3)_2$, 303 K, -750 mV] and Co+PANI [$Co(NO_3)_2$ with polyaniline-derivative, 303 K, -750 mV].

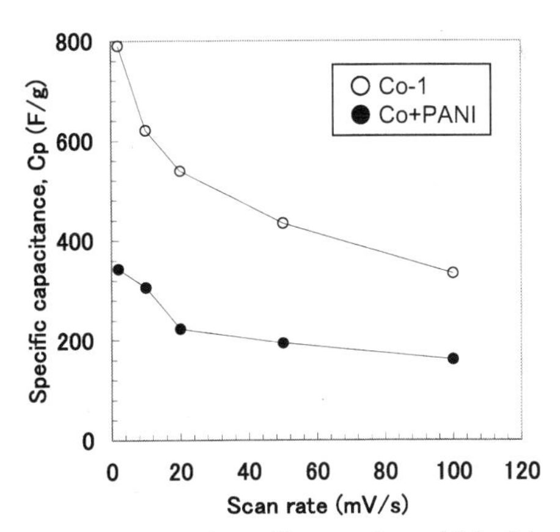

Figure 6 Relationship between scan rate and specific capacitance (Cp) of deposit films of Co-1 sample [deposition condition: $Co(NO_3)_2$, 303 K, -750 mV] and Co+PANI [$Co(NO_3)_2$ with polyaniline-derivative, 303 K, -750 mV].

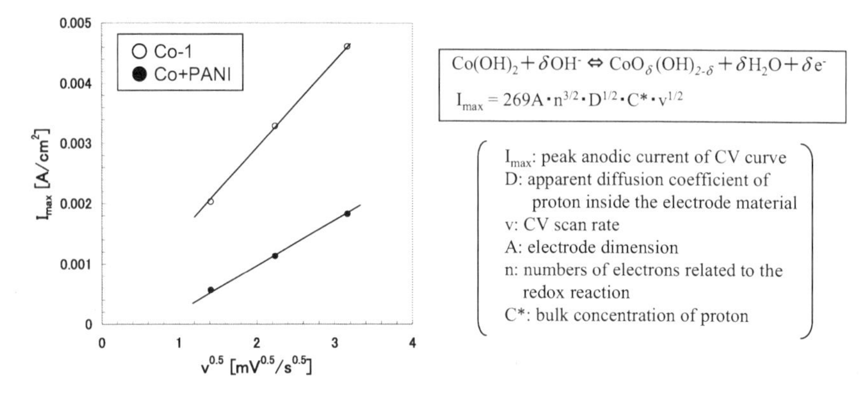

Figure 7 Relationship between scan rate and observed anodic peak current value of CV curve for deposit films: Co-1 sample [deposition condition: $Co(NO_3)_2$, 303 K, -750 mV] and Co+PANI [$Co(NO_3)_2$ with polyaniline-derivative 303 K, -750 mV]; the slope of the line are proportional to $D^{1/2}$ (D: apparent diffusion coefficient of proton inside the electrode material).

CONCLUSIONS

We prepared α- and β-$Co(OH)_2$ films were prepared by the electrolytic deposition using $Co(NO_3)_2$ aqueous solution. Alpha-$Co(OH)_2$ films were formed at 303K and β-$Co(OH)_2$ films were found to form preferably rather than α-$Co(OH)_2$ at the increased temperature (348 or 363 K). The films with scale-like α-$Co(OH)_2$ particles, which are expected to be poorly crystalline, showed high specific capacitance (Cp) value of 790 F/g. On the contrary, β-$Co(OH)_2$ films with thicker particles showed low Cp value of 41 F/g. Their crystallinity and morphologies would significantly affect their Cp values.

The α-$Co(OH)_2$/polyaniline composite film was prepared using $Co(NO_3)_2$ aqueous solution with the water-soluble conducting polymer (polyaniline-derivative). The film was composed of diminished sized α-$Co(OH)_2$ particles and polyaniline-derivative. The Cp value of the composite film was only about half of the α-$Co(OH)_2$ film. The Cp values were much lower than expected for the materials with increased electrode-electrolyte interface area and with co-existing polyaniline-derivative. It was indicated that the proton diffusion inside the electrode active material for the composite film was lower than that of α-$Co(OH)_2$ film. The possible interpretation is that the coexisting polyaniline-derivative would cause the decreased proton concentration and insufficient electrical conductivity for the composite film. In reducing ambient (0.1 M KOH electrolyte), even for the self-doped polyaniline, its electrical conductivity would be much depressed. It is suggested that the morphology control giving best balance between proton diffusion and electron conduction should be necessary in order for further enhancing Cp values.

REFERENCES

[1]Y. Zhang, H. Feng, X.B. Wu, L.Z. Wang, A.Q. Zhang, T.C. Xia, H.C. Dong, X.F. Li and L.S. Zhang, Progress of Electrochemical Capacitor Electrode Materials: A Review, *Int. J. Hydrogen Energy*, **34**, 4889-4899 (2009).

[2]J.P. Zheng, P.J. Cygan and T.R. Jow, Hydrous Ruthenium Oxide as an Electrode Material for Electrochemical Capacitors, *J. Electrochem.Soc.*, **142**, 2699-2703 (1995).

[3]V. Gupta, T. Kusahara, H. Toyama, S. Gupta and N. Miura, Potentiostatically Deposited Nanostructured α-$Co(OH)_2$: A High Performance Electrode Material for Redox-capacitors, *Electrochem. Commun.*, **9**, 2315-2319 (2007).

[4]W.-J. Zhou, J. Zhang, T. Xue, D.-D. Zhao and H.-L. Li, Electrodeposition of Ordered Mesoporous Cobalt Hydroxide Film from Lyotropic Liquid Crystal Media for Electrochemical Capacitors, *J. Mater. Chem.*, **18**, 905-910 (2008).

[5]M.S. Yarger, E. Steinmiller and K.-S. Choi, Electrochemical Synthesis of Cobalt Hydroxide Films with Tunable Interlayer Spacings, *Chem. Commun.*, 159-161 (2007).

[6]M. Kurmoo, Ferrimagnetism in Dicarboxylate-Bridged Cobalt Hydroxide Layers, *J. Mater. Chem.*, **9**, 2595-2598 (1999).

ACKNOWLEDGMENT

Authors thank Mr. Juntaro Seki and Mr. Hiroaki Kadoura for SEM observations, Mr. Yusuke Akimoto for TEM observations, Mr. Yasuhito Kondo and Dr. Ryoji Asahi of Toyota Central R&D Labs., Inc. for fruitful discussion on the electrolytic depositions.

OPTIMIZATION OF SPARK-PLASMA-SINTERING CONDITIONS FOR MAXIMIZING FIGURE OF MERIT OF La-DOPED $SrTiO_3$

Asami Kikuchi[a], Sidney Lin[b], Noriyuki Okinaka[c] and Tomohiro Akiyama[c]
[a]Graduate School of Engineering, Hokkaido University
Sapporo, Hokkaido, Japan
[b]Dan F. Smith Department of Chemical Engineering, Lamar University
Beaumont, Texas, USA
[c]Faculty of Engineering, Hokkaido University
Sapporo, Hokkaido, Japan

ABSTRACT

La-doped $SrTiO_3$ was prepared by a new method; combustion synthesis (CS) followed by a spark plasma sintering (SPS). The figures of merits of prepared samples were measured to study the effect of sintering temperature and holding time on the thermoelectric performance. $Sr_{0.92}La_{0.08}TiO_3$ powders were prepared by CS from oxides, carbonates, metals, and sodium perchlorate. SPS was carried out in vacuum at different sintering temperatures ranged from 1573 to 1663K for different holding times for 1, 5, 15, and 30 min to decide the optimum sintering conditions. Our results show the sample sintered at 1573K for 5 min reaches the maximum *ZT* value of 0.37 at 1045 K. This is the largest ZT value reported thus far for $SrTiO_3$-based bulk semiconductors as far as we know. The results also suggest that the combination of CS with post SPS can be used to commercially produce thermoelectric oxides for use in efficient electricity generation.

INTRODUCTION

Many researchers have studied the thermoelectric properties of $SrTiO_3$-based materials[1-6]. There have been many investigations on the fabrication methods of $SrTiO_3$ ceramics, namely, the wet-chemical method[7], the solid-state reaction method[8], and the sol–gel method[9], which are used in the production of many materials. However, these conventional methods are quite time- and energy-consuming. To overcome this problem, we have been focused on a combination of combustion synthesis (CS)[10] and spark plasma sintering (SPS)[6,11-19]. This proposed method is suitable for synthesizing thermoelectric materials and has many benefits. SPS is one of the sintering methods based on the spark plasma phenomenon generated by the high charging currents between particles. As compared to hot-press sintering method, SPS has a very high heating rate and mass transfer rate, enabling samples to be densely sintered in a short period at a relatively low temperature. Thus, the grain growth in the sample might be suppressed, leading to low thermal conductivity, which means that SPS appears to be more suitable for preparing thermoelectric materials[17]. It is widely known that sintering conditions such as sintering temperature[20-22] and holding time during sintering seriously affect the microstructures and properties of materials. However, the effect of spark-plasma-sintering conditions on the thermoelectric properties has not yet been satisfactorily explained. Therefore, the purpose of this study is to investigate figures of merit of La-doped $SrTiO_3$ prepared by CS with post SPS, with the mainly focused on the study of the effect of spark-plasma-sintering conditions. The new findings of this study provide information for maximizing the figures of merit not only of $SrTiO_3$-based materials but also of various other oxide-based materials.

EXPERIMENTS

Polycrystalline samples of $Sr_{0.92}La_{0.08}TiO_3$ (SLTO) were prepared from $SrCO_3$ (99.9% purity, Kanto Chemical Co. Inc., Tokyo, Japan), TiO_2 (99.9% purity, Kanto Chemical Co. Ltd., Sakado, Japan), Ti (99.9% purity, Kojundo Chemical Co. Ltd., Sakado, Japan), $NaClO_4$ (98.0% purity, Sigma–Aldrich, St. Louis, MO, USA), and La_2O_3 (99.9% purity, Kojundo Chemical

Laboratory Co. Ltd., Sakado, Japan). Details of the experimental procedure can be found from our previous publications [13,14]. The combustion-synthesized (CSed) products were pulverized into powders in a planetary ball mill (Pulverisette 6, Fritsch, Idar-Oberstein, Germany) operated at 350 rpm for 40 min in air. The average particle size of the ground powders was measured using a fiber optic dynamic light-scattering photometer (Otsuka Electronics Co. Ltd., Tokyo, Japan). The powders obtained were placed in a cylindrical graphite die and sintered by SPS (SPS-511S, Sumitomo Coal Mining, Tokyo, Japan) at a heating rate of 30 $Kmin^{-1}$ in a vacuum under mechanical pressure conditions of 34 MPa produced using plungers. Experiments to find the optimum sintering conditions were taken part in 2 stages. At the first stage, we sintered samples by changing sintering temperatures from 1573 to 1663K in 30K apart for a fixed time of 15 min. At the second stage, samples were sintered for different holding times from 1 to 30 min at an obtained temperature at the first stage. The phase compositions and morphologies of the products were analyzed using an X-ray diffractometer (Miniflex, Rigaku, Tokyo, Japan) and a scanning electron microscope (SEM) (JSM- 7000F, JEOL, Tokyo, Japan). The electrical resistivities and Seebeck coefficients were measured simultaneously using a Seebeck coefficient/electric resistance measuring system (ZEM-3, ULVAC-RIKO, Yokohama, Japan) from room temperature to 1045 K in helium atmosphere. The thermal conductivity was calculated as

$$\kappa = DC_p d \tag{1}$$

where κ, D, Cp, and d are the thermal conductivity, thermal diffusivity, heat capacity, and experimental density, respectively. The densities of the samples were measured by the Archimedes method, and the thermal diffusivities and heat capacities were measured using a laser flash thermal constant analyzer (TC-7000, ULVAC-RIKO, Yokohama, Japan) from room temperature to 1105 K in vacuum. The oxygen defect content was determined by measuring the weight increase of the sample after heating at 1373 K for 24 h in air and were completely oxidized to $Sr_{0.92}La_{0.08}TiO_{3.00}$. In a preliminary test, sintered SLTOs (10.0 mg) were heated to 1373 K at a heating rate of 3 $Kmin^{-1}$ and an air-flow rate of 50 $mlmin^{-1}$. The weight of the sample gradually increased and then it reached a steady state after 12 h, therefore an experimental time of 24 h was enough for perfect oxidation of the sample.

RESULTS AND DISCUSSION

Reaction analyses of CSed SLTO

The CS reaction equation is given as follows[13, 14]:

$$(1-x)SrCO_3 + (1-a)Ti + aTiO_2 + \frac{x}{2}La_2O_3 + \frac{4-4a-x}{8}NaClO_4$$
$$\longrightarrow Sr_{1-x}La_xTiO_3 + (1-x)CO_2 + \frac{4-4a-x}{8}NaCl. \tag{2}$$

In Eq. (2), x denotes the La-doping content and a denotes the TiO_2 content. In this study, x and a are equal to 0.08[11] and 0.25[10], respectively. Figure 1 shows XRD patterns of samples before and after CS. All peaks after CS very well correspond to that of $SrTiO_3$. This result indicates that SLTO with high purity was prepared by CS.

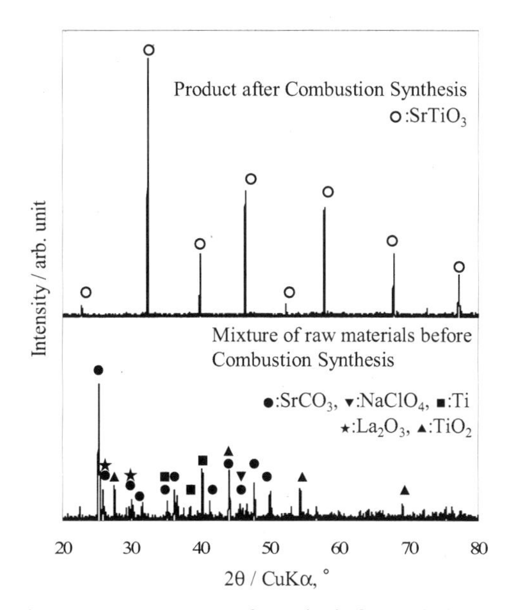

Figure 1. XRD patterns of samples before and after CS.

Stage 1 – The effect of sintering temperature on thermoelectric properties of SLTO

Table I lists sintering temperatures and bulk densities after SPS. The values of for sample sintered at various sintering temperatures are also shown in this table. As shown in Table I, density of all samples reached higher than 96.3% of the theoretical density (T.D.) (T.D. = 5.24 gcm^{-3}). Experiments were conducted at sintering temperatures range from 1513 to 1663K, however bulk SLTO was not formed at a sintering temperature of 1663 K, because of the formation of many intercrystalline cracks. Thus, we concluded that sintering of SLTO was successfully carried out at temperatures from 1513 to 1633 K.

Table I Sintering temperatures, bulk densities after SPS and values of (oxidation defect content).

Sintering temperature [K]	Bulk density after SPS ($g\cdot cm^{-3}$)	(%T.D.)	The value of δ [–]
1513	5.03	96.3	0.01
1543	5.15	98.6	0.03
1573	5.14	98.5	0.03
1603	5.09	97.6	0.03
1633	5.16	98.9	0.05
1663	5.09	97.6	–

T.D. = 5.24 $g\cdot cm^{-3}$

Figures 2(a), (b), (c) and (d)[13] show the temperature dependence of electric conductivity, Seebeck coefficient, thermal conductivity and dimensionless figure of merit of CSed and SPSed

SLTO for various sintering temperatures, respectively. As temperature increased electric conductivity decreased and the absolute value of Seebeck coefficient increased, indicating a metallic behavior. In the experimental temperature range, with an increase in the sintering temperature electric conductivity increased and the absolute value of Seebeck coefficient decreased due to an increase of the carrier electrons. When a sample was sintered at high temperature in abutting contact with carbon in vacuum, a lot of oxygen inside particles was liberated. As a result, the sample changed into the nonstoichiometric compound of $Sr_{0.92}La_{0.08}TiO_{3-}$, where is the oxygen defect content. Oxygen defects generate electrons as carriers. As shown in Table I, the largest value of is 0.05 in a sample sintered at 1633 K, and the smallest value of is 0.01 in a sample sintered at 1513 K. The value of of the samples sintered at different temperatures shows a similar tendency with electric conductivity of the samples. It can been seen in Figure 2(a) that electric conductivity of samples sintered at 1543, 1573, and 1603K were almost the same so were their values in Table I. Thus, an increase in the sintering temperature leads to an increase in the oxygen defect content, that is to say, an increase in carrier electrons, which results in a higher electric conductivity. Therefore, electric conductivity increased and the absolute value of Seebeck coefficient decreased with an increase in the sintering temperature.

Thermal conductivity of CSed and SPSed SLTO increased with sintering temperature. With an increase in sintering temperature, average grain size of CSed and SPSed SLTO increased[13], which caused the decrease of the effects of phonon scattering at grain boundaries. Therefore, thermal conductivity of CSed and SPSed SLTO increased with an increase in sintering temperature. ZT of samples sintered at more than 1543K showed peaks at around 1000 K. ZT of sample sintered at 1513K showed very low values due to its low electric conductivity shown in Figure 2(a). Among the CSed and SPSed SLTO, sample sintered at 1573K showed the maximum ZT of

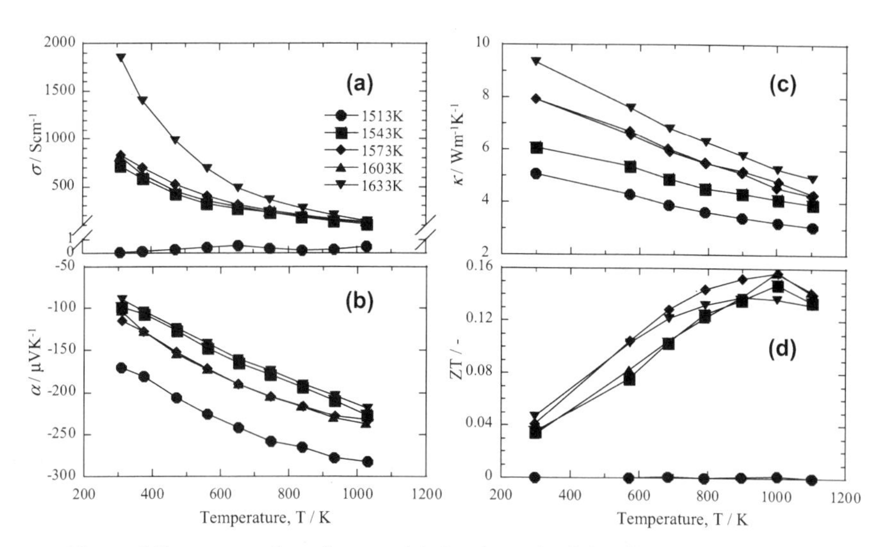

Figures 2 Temperature dependence on (a) electric conductivity, (b) Seebeck coefficient, (c) thermal conductivity and (d) dimensionless figure of merit of CSed and SPSed SLTO for various sintering temperatures.

0.16 at 1005 K. These results indicate that for production of SLTO by the combination of CS and SPS, sintering should be done at 1573 K.

Stage 2 – The effect of sintering time on thermoelectric properties of SLTO

Table II lists holding times, bulk densities after SPS and values of for samples sintered for various holding times. The densities of the samples sintered for more than 5 min reached 96.8% of T.D..The density of the sample sintered for 1 min reached only 66.2% of T.D., which was 46.2% lower than in the other cases. This showed that bulk SLTO was densely sintered when the sintering times were greater than 5 min.

Table II Holding times, bulk densities after SPS and values of (oxidation defect content).

Holding time [min]	Bulk density after SPS ($g\cdot cm^{-3}$)	(%T.D.)	The value of δ [–]
1	3.47	66.2	0.02
5	5.12	97.7	0.03
15	5.14	98.5	0.03
30	5.07	96.8	0.06

T.D. = 5.24 $g\cdot cm^{-3}$

Figures 3 (a), (b), (c) and (d)[14] show the temperature dependence of the electric conductivities, the Seebeck coefficients, the thermal conductivities and dimensionless figure of merit of the products obtained by sintering for different holding times. With increasing sintering times, the electrical conductivities increased and the absolute values of the Seebeck coefficient

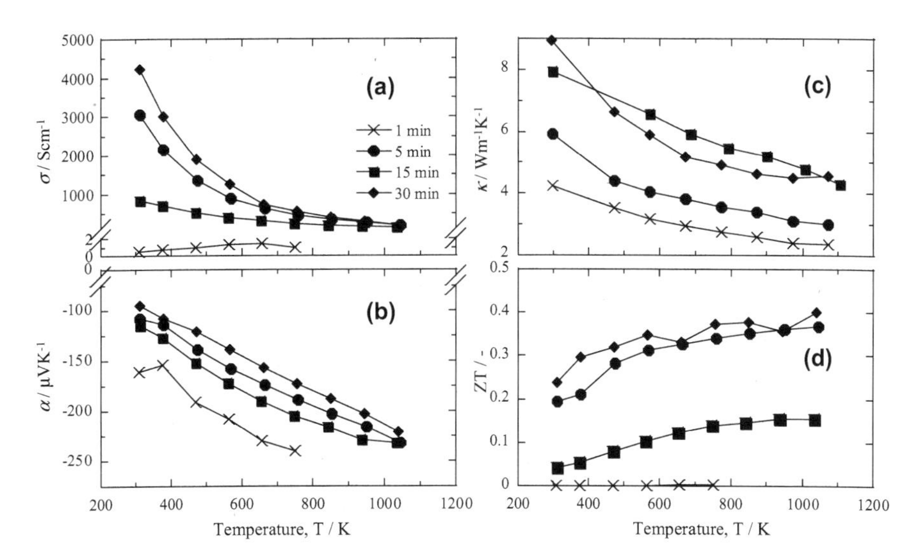

Figures 3 Temperature dependence on (a) electric conductivity, (b) Seebeck coefficient, (c) thermal conductivity and (d) dimensionless figure of merit of CSed and SPSed SLTO for various holding times.

decreased, except the 5-min-sintered sample. This was caused by reduction of SLTO powders during SPS as mentioned in the section of stage 1.

In the temperature range considered, ZT values of the sample sintered for 1 min were very low because of the sample's low electric conductivity, as shown in Figure 3 (a). The ZT values of the 5-min-sintered sample increased with temperature, and the values have a tendency to increase at even higher temperatures, which showed that this sample is suitable for high-temperature applications. Among our samples, SLTO sintered for 5 min showed the maximum ZT of 0.37 at 1045 K. This value is the largest value ever reported for $SrTiO_3$-based bulk semiconductors, apart from that of Nb-doped $SrTiO_3$ epitaxial films reported by Ohta et al.[4]. Generally, thermoelectric modules are composed of bulk polycrystalline materials. Our sample can be used for this application because of its high ZT values and suitability for high-temperature usage. The optimum holding time for sintering was 5 min at 1573 K[13,14]. All of these results proved that CS with post SPS is a promising.

CONCLUSIONS

We studied the effect of sintering conditions on figures of merit of La-doped $SrTiO_3$ prepared by a combination of CS with post SPS. The following results were obtained: (1) Spark-plasma-sintering of $Sr_{0.92}La_{0.08}TiO_3$ was successfully carried out at 1513 to 1633K in temperature. Sintering at a temperature above 1633K didn't form, $Sr_{0.92}La_{0.08}TiO_3$ bulk due to the formation of many intercrystalline cracks. (2) High-temperature sintering resulted in oxygen deficiency which produced electrons as carriers. Thus, samples sintered at a higher temperature showed larger electric conductivity. (3) The optimum sintering temperature to maximize ZT value was 1573K. (4) Prolonged sintering introduced more oxygen defects into the sample, which can produce carrier electrons. Thus, samples sintered for longer times, except for the sample sintered for 15 min, showed higher electric conductivities. (5) The optimum holding time during sintering maximized ZT was 5 min when sintering temperature was 1573K. (6) The sample sintered for only 5 min recorded the maximum ZT value of 0.37 at 1045K. This is the largest value reported thus far for $SrTiO_3$-based bulk semiconductors. These results also suggest that the combination of CS with post SPS can be used to commercially produce thermoelectric oxides for use in efficient electricity generation.

REFERENCES

[1]H. Muta, K. Kurosaki, and S. Yamanaka, Thermoelectric properties of rare earth doped $SrTiO_3$, *J. Alloy. Compd.,* **350**, 292-295 (2003).

[2]H. Obara, A. Yamamoto, C. H. Lee, K. Kobayashi, A. Matsumoto, and R. Funahashi, Thermoelectric Properties of Y-Doped Polycrystalline $SrTiO_3$, *Jpn. J. Appl. Phys.,* **43**, 540-542 (2004).

[3]T. Okuda, K. Nakanishi, and K. Yamanaka, Large thermoelectric response of metallic perovskites: $Sr_{1-x}La_xTiO_3$ ($0<x<0.1$), *Phys. Rev. B Mat.,* **63**, 113104-113101 (2001).

[4]S. Ohta, T. Nomura, H. Ohta, M. Hirano, H. Hosono, and K. Koumoto, Large thermoelectric performance of heavily Nb-doped $SrTiO_3$ epitaxial film at high temperature, *Jpn. J. Appl. Phys.,* **87**, 092108-092101-092103 (2005).

[5]S. Ohta, T. Nomura, H. Ohta, and K. Koumoto, High-temperature carrier transport and thermoelectric properties of heavily La- or Nb-doped $SrTiO_3$ single crystals, *Jpn. J. Appl. Phys.,* **97**, 034106-1–034106-4 (2005).

[6]I. J. Shon, Z. A. Munir, K. Yamazaki, and K. Shoda, Simultaneous Synthesis and Densification of $MoSi_2$ by Field-Activated Combustion, *J. Am. Ceram. Soc.,* **79**, 1875-1880 (1996).

[7]P. Balaya, M. Ahrens, M. L. Kienle, J. Maier, B. Rahmati, S. B. Lee, and W. Sigle, Synthesis and characterization of nanocrystalline $SrTiO_3$, *J. Am. Ceram. Soc.*, **89**, 2804 (2006).
[8]D. Cadavid and J. E. Rodrìguez, Thermoelectric properties of polycrystalline Zn_4Sb_3 samples prepared by solid state reaction method, *Physica B*, **403**, 3976-3979 (2008).
[9]K. F. Cai, E. Müller, C. Drašar, and A. Mrotzek, Preparation and thermoelectric properties of Al-doped ZnO ceramics, *Mater. Sci. Eng. B*, **104**, 45-48 (2003).
[10]H. Ishikawa, K. Oohira, T. Nakajima, and T. Akiyama, Combustion synthesis of $SrTiO_3$ using different raw materials, *J. Alloy. Compd.*, **454**, 384-388 (2008).
[11]L. Zhang, T. Tosho, N. Okinaka, and T. Akiyama, Thermoelectric Properties of Combustion-Synthesized Lanthanum-Doped Strontium Titanate, *Mater. Trans.*, **48**, 1079-1083 (2007).
[12]L. Zhang, T. Tosho, N. Okinaka, and T. Akiyama, Thermoelectric Properties of Combustion Synthesized and Spark Plasma Sintered $Sr_{1-x}R_xTiO_3$ (R=Y, La, Sm, Gd, Dy, $0 \le x \le 0.1$), *Mater. Trans.*, **48**, 2088-2093 (2007).
[13]A. Kikuchi, L. Zhang, N. Okinaka, T. Tosho, and T. Akiyama, Optimization of Sintering Temperature for Maximizing Dimensionless Figure of Merit of La-Doped Strontium Titanate Thermoelectric Material in the Combination of Combustion Synthesis with Post Spark Plasma Sintering, *Mater. Trans.*, **51**, 1919-1922 (2010).
[14]A. Kikuchi, N. Okinaka, and T. Akiyama, A large thermoelectric figure of merit of La-doped SrTiO3 prepared by combustion synthesis with post-spark plasma sintering, *Scripta Mater.*, **63**, 407-410 (2010).
[15]T. Nishimura, M. Mitomo, H. Hirotsuru, and M. Kawahara, Fabrication of silicon nitride nano-ceramics by spark plasma sintering, *J. Mater. Sci. Lett.*, **14**, 1046-1047 (1995).
[16]H. Simizu, M. Yoshinaka, and K. Hirota, Fabrication and mechanical properties of monolithic $MoSi_2$ by spark plasma sintering, *Mater. Res. Bull.*, **37**, 1557-1563 (2002).
[17]Y.-H. Lin, Z. Shi, C.-W. Nan, Y. Liu, and J. Li, High-Temperature Electrical Transport and Thermoelectric Power of Partially Substituted $Ca_3Co_4O_9$-Based Ceramics, *J. Am. Ceram. Soc.*, **90**, 132-136 (2007).
[18]Y. Liu, Y. Lin, Z. Shi, C.-W. Nan, and Z. Shen, Preparation of $Ca_3Co_4O_9$ and Improvement of its Thermoelectric Properties by Spark Plasma Sintering, *J. Am. Ceram. Soc.*, **88**, 1337-1340 (2005).
[19]J. J. Petrociv and R. E. Honnell, Partially stabilized ZrO_2 particle-$MoSi_2$ matrix composites, *J. Mater. Sci.*, **25**, 4453-4456 (1990).
[20]R.-C. Chang, S.-Y. Chu, Y.-P. Wong, C.-S. Hong, and H.-H. Huang, The effect of sintering temperature on the properties of lead-free $(Na_{0.5}K_{0.5})NO_3-SrTiO_3$ ceramics," *J. Alloy. Compd.*, **456**, 308-312 (2008).
[21]X. Dong, F. Lti, L. Yang, Y. Zhang, and X. Wang, Influence of spark plasma sintering temperature on electrochemical performance of $La_{0.80}Mg_{0.20}Ni_{3.75}$ alloy, *Mater. Chem. Phys.*, **112**, 596-602 (2008).
[22]Q. Hu, P. Luo, and Y. Yao, Influence of spark plasma sintering temperature on sintering behavior and microstructure of bulk $MoSi_2$, *J. Alloy. Compd.*, **459**, 163-168 (2008).

Advanced Materials and Technologies for Rechargeable Batteries

DESIGN OF (THIO) PHOSPHATES FOR HIGH PERFORMANCE LITHIUM ION BATTERIES

Stefan Adams, Rayavarapu Prasada Rao
National University of Singapore, Department of Materials Science and Engineering
Singapore 117576, SINGAPORE

ABSTRACT

Empirical bond length - bond valence relations provide insight into the link between structure and ion transport in solid electrolytes and mixed conductors. Building on our earlier systematic adjustment of bond valence (BV) parameters to the bond softness, here we link the squared BV mismatch to the absolute energy scale and use it as a Morse-type interaction potential for analyzing low-energy ion migration paths in ion or mixed-conducting solids by either an energy landscape approach or molecular dynamics (MD) simulations and compare the results to experimental characterizations. For a wide range of lithium oxide and lithium sulfide compounds we could thus model ion migration pathways and mechanisms revealing significant differences to an earlier geometric approach. This novel BV-based force-field has then been applied to investigate a range of mixed conductors, focusing on cathode materials for lithium ion battery (LIB) applications to promote a systematic design of LIB cathodes that combine high energy density with high power density. To demonstrate the versatility of the new BV-based force field it is applied in exploring various strategies to enhance the power performance of phosphate-based safe low-cost cathode materials including $LiFePO_4$, $LiVPO_4F$ and thiophosphate solid electrolytes for rechargeable all-solid-state lithium Li-ion batteries (AS-LIBs). The argyrodite-type Li_6PS_5X (X = Cl, Br, I) thiophosphates were prepared by mechanical milling and subsequent annealing. Samples are characterized structurally by neutron and X-ray powder diffraction as well as electrochemically by impedance spectroscopy. A room temperature conductivity of $10^{-3.1}$ S/cm renders the anion-disordered Li_6PS_5X solid electrolytes (with X=Cl, Br) suitable for AS-LIBs.

INTRODUCTION

Understanding ionic motion in disordered solids requires insight into the correlation between ion mobility and the structural and energetic environment of mobile ions. Local structure models for disordered solid electrolytes such as ion conducting glasses may in principle be derived from diffraction data (via crystal structure refinements or in the case of glasses by reverse Monte Carlo (RMC) fits) or molecular dynamics (MD) simulations.[1,2] MD simulations yield comprehensive structural and dynamical information within the limitations imposed by the size of the simulated system, the time span covered by the simulation and the agreement of the employed interaction potential parameters with reality. Both approaches proved to be valuable tools in obtaining insights into conduction mechanisms and their correlation to the atomic structure, though in the case of MD simulations it has to be verified that the force field chosen for the simulations leads to structure models that are consistent with experimental information.[2-4] Here, we discuss how the bond valence (BV) method can be used to predict characteristics of ionic conductivity from structure models,[4-7] and be optimized by linking it to an absolute energy scale for a more straightforward comparison with other simulation tools. Computational studies are complemented by experimental structural and electrochemical characterizations.

COMPUTATIONAL METHODS

Bond valence pathway models

Empirical relationships between bond length R and bond valence $s_{A-X} = exp[(R_0-R)/b]$ are widely used in crystal chemistry to identify plausible equilibrium sites for an atom in a structure as sites, where the BV sum of the atom matches its oxidation state (*see e.g.* the recent review by Brown[5] and references therein). In our earlier work, we introduced a systematic adjustment of the BV parameter b to bond softness,[6-9] which together with the inclusion of interactions beyond the first coordination shell permits more adequate estimates of non-equilibrium site energies. The inclusion of weak interactions to counter-ions beyond the first coordination shell is indispensable for modelling ion transport to avoid artefacts when an ion moves across the border of its coordination shell. Low BV sum mismatch (and hence low energy) pathways for a cation A^+ can then be modelled as regions in the structure, where the BV sum $V(A)= \Sigma s_{A-X}$ (summing up over all adjacent counter-ions X) approaches the "ideal valence" $V_{id}(A)$, *i.e.* which is identified with the magnitude of its oxidation state.

To enhance chemical plausibility of such "BV mismatch landscapes" | $V(A)$|, penalty functions p_{A-X} have been introduced that (i) discriminate against sites, where a matching $V(A)$ is achieved by strongly asymmetric coordinations (for details see refs. 3, 6) and (ii) exclude sites close to other (immobile) cation types. These cation-cation penalty functions may simply take the form of exclusion radii, but truncated Coulomb repulsions yield a more physical description. While it is obvious that a higher BV mismatch implies an energetically less favourable state, a direct link of the type

$$E(A)= D_0 \cdot \left| \sum_X s_{A-X} - V_{id}(A) \right|^g + E_{asym} + E_{rep} \tag{1}$$

remained to be established.[4] In eq. (1), E_{asym} represents an energy penalty term due to the asymmetry of the coordination of the considered ion and E_{rep} the (Coulomb) repulsion between mobile and immobile cations.

Bond-valence based force-field

An empirical approach to assess the coefficients D_0 and g in eq. (1) and a suitable functional form for E_{asym} may start from comparing the distance dependence of the BV sum mismatch and of the interaction energy in empirical interatomic potentials such as a Morse-type potential

$$E = D_0 \left\{ (\exp[\alpha(R_{\min} - R)] - 1)^2 - 1 \right\} = D_0 \left\{ \frac{1}{s_{\min}^2} \left(\exp\left[\frac{R_0 - R}{b} \right] - s_{\min} \right)^2 - 1 \right\} \tag{2}$$

with $b = 1/\alpha$. As seen, the interaction energy E can be approximated as a quadratic function (i.e. g=2 in eq. 1) of the deviation of the BV from its value $s_{min} = exp[(R_0 - R_{min})/b]$ for the energy minimum distance ($R = R_{min}$). Note that the BV parameter R_0 (the distance corresponding to $s = 1$) generally differs from R_{min}. Introducing the relative bond valence $s_{rel} = s/s_{min}$, eq. 2 can then be expressed concisely as:[4]

$$E = D_0 \left\{ \frac{(s - s_{\min})^2}{s_{\min}^2} - 1 \right\} = D_0 \left(s_{rel}^2 - 2 s_{rel} \right) \tag{3}$$

The Morse-type interaction potential in eq. (2) or (3) is characterised by the three parameters D_0, R_{min} and α that have to be linked to the BV parameters. As mentioned above, α is simply $1/b$. A tentative approach to establish a consistent set of R_{min} (or s_{min}) values may be expressed as

$$R_{\min} \approx R_0 \times \left[0.9185 + 0.2285 \cdot \left| \sigma_A - \sigma_X \right| \right] - b \cdot \ln\left(\frac{V_{id}}{N_C} \right) \tag{4}$$

where N_C refers to the preferred coordination number of the central ion (c.f. Table 1) and the empirically determined term in square brackets accounts for the effect of polarisation (σ_A, σ_X refer to the absolute softnesses of the cation and anion, respectively; see *e.g.*, ref. 8) as well as the influence of higher coordination shells.

The dissociation energy D_0 can be expected to be $D_0 = b^2\ k/2$, k being the force constant for $R = R_{min}$. We have thus approximated D_0 for a wide range of cations as

$$D_0 = \frac{k \cdot b^2}{2} = c \cdot 14.4 \frac{eV}{\AA} \cdot \frac{\left(V_{id}(A) \cdot V_{id}(X)\right)^{1/c}}{R_{\min}\sqrt{n_A \cdot n_X}} \times \frac{b^2}{2} \tag{5}$$

with $c = 1$ if A is an s or p block elements, or $c = 2$ if A is a d or f block element. n_A, n_X represent the principal quantum numbers of cation A and anion X and $V_{id}(A)$, $V_{id}(X)$ the absolute value of their respective nominal charges.

In contrast to the BV sum mismatch description, such BV interaction potentials fulfil the formal requirements for an anharmonic diatomic interaction potential, allowing for molecular dynamics simulations. Due to the ready availability of BV parameters for a wide range of cation anion pairs this might be an attractive option. In table 1 we list parameters derived in this way from our respective *softBV* bond valence parameters[8, 9] for 132 cation types in oxides. With this definition of the bond energy, the total site energy $E(A)$ of a cation A, can then be determined as the sum over terms for the interactions with each of the N adjacent anions X_i :

$$E_{pot}(A) = D_0 \left[\sum_{j=1}^{N} \left(\frac{s_{A-Xj} - s_{\min}}{s_{\min}} \right)^2 - N \right] \tag{6}$$

By rewriting eq.(6), it becomes obvious that the total potential energy varies with both the mismatch of the BV sum and the asymmetry of the coordination. This permits a quantification of the correlation between:

1. the *bond valence sum rule,*[10] stating that the sum of bond valences around an atom equals its atomic valence;
2. the *equal valence rule,*[10] which states that the sum of the bond valences around any loop is zero, *i.e.* the most symmetric distribution of atomic valence among the bonds is energetically preferable.

If for the sake of simplicity only contributions from the N_C counterions of type X in the 1st coordination shell of A are considered in the derivation, the correlation takes the simple form of eq. (7):

$$E_{pot}(A) = D_0 \left[N_C \cdot \left\{ \left(\frac{V(A) - V_{\min}(A)}{V_{\min}(A)} \right)^2 - 1 \right\} + \sum_{i=1}^{N_C} \left(\frac{s_{A-X} - \bar{s}_{A-X}}{s_{\min}} \right)^2 \right] \tag{7}$$

where $V_{min}(A) = N_C \times s_{min}$ in the first (BV sum mismatch) term , while the second (asymmetry) term quantifies the effect of the deviation of individual bond valences from their average value $\bar{s}_{A-X} = V(A)/N_C$.[4] A major advantage of such an energy-scaled BV mismatch is that it allows straightforward combinations of the BV-based "attraction term" with suitably weighted penalty functions for coordination asymmetry and particularly a Coulombic repulsion. Fractional charges from first principles might improve the quality of the simulation, yet prevent a fast and automatic generation of force-fields for screening of a wide range of compounds. Thus we estimated the q_A, q_X. simply from

$$q_{Ai} = \frac{V_{id}(A_i)}{\sqrt{n_{Ai}}}\sqrt{\frac{\sum_j \frac{V_{id}(X_j)\cdot N_{Xj}}{\sqrt{n_{Xj}}}}{\sum_i \frac{V_{id}(A_i)\cdot N_{Ai}}{\sqrt{n_{Ai}}}}},\ q_{Xj} = \frac{V_{id}(X_j)}{\sqrt{n_{Xj}}}\sqrt{\frac{\sum_i \frac{V_{id}(A_i)\cdot N_{Ai}}{\sqrt{n_{Ai}}}}{\sum_j \frac{V_{id}(X_j)\cdot N_{Xj}}{\sqrt{n_{Xj}}}}} \tag{8}$$

in which N_{Ai} (N_{Xj}) refers to the occupancies of the i-th cation A_i (j-th anion X_j) in the structure ensuring electroneutrality. Coulomb repulsions e.g. between immobile A_1 and mobile A_2 cations are then taken into account in a screened version $E_{Coulomb}(A_1\text{-}A_2)$:

$$E_{Coulomb}(A_1 - A_2) = \frac{q_{A1}\cdot q_{A2}}{R_{A1-A2}}\cdot erfc\left(\frac{R_{A1-A2}}{\rho_{A1-A2}}\right) \tag{9}$$

The screening factor $\boldsymbol{\rho_{A1\text{-}A2} = (r_{A1} + r_{A2})\,f}$ is assumed to equal the sum of the covalent radii $\boldsymbol{r_{Ai}}$ of the two ions involved times a factor $\boldsymbol{f}$ that depends on the average absolute cation electronegativity and the average cation charge in the compound. Typical values of $\boldsymbol{f}$ in ternary and quaternary lithium oxides fall into the range 0.74 ± 0.04 and thereby $\boldsymbol{\rho}$ is of the order of 2 Å. This simplification restricts long range interactions to the real part of the Ewald sum, but such a localized interaction model has been shown to lead to realistic activation energies of diffusion e.g. for a range of Li conducting oxyacid salts.[11-13] Due to the favourable convergence of the interaction model a relatively short cut-off radius (8-10 Å) could be generally used enhancing computational efficiency.

EXPERIMENTAL METHODS

$LiVPO_4F$ was prepared by mixing stoichiometric amounts of VPO_4 and LiF (Merck, 99 % purity) using a high energy ball mixer (Spex Ball mixer (D8000), USA) for 12 h (speed: 1400rpm, Ball-powder ratio (in g) : 10:2.5) and heating the mixture at 700 °C for 1h in Ar atmosphere. VPO_4 was prepared by carbothermal reduction method similar to Barker et al.[14], reacting V_2O_5 (5.42 g, Aldrich; purity, 98%) and ammonium dihydrogen phosphate ($NH_4H_2PO_4$) (3.93 g, Merck; purity, 99%)) and carbon (0.43 g, Super P, surface area: 250 m^2 / g, 20 wt % excess carbon), which were mixed using a mechanical grinder, pelletized and heated at 750 °C for 4 h, in flowing Ar-gas in a tubular furnace.

Li_6PS_5X (X= Cl, Br, I) samples were prepared by high energy mechanical milling using Agate (45 ml pot and 15 number of 10 mm ø ball) pot and balls; Li_2S, P_2S_5 and LiX (X= Cl, Br, I) crystalline powders were used as starting materials. The samples were ground for 20h followed by annealing at 550°C for 5 h. Samples were pressed into 10 mm diameter pellets of ~ 1.5 mm thick for annealing the samples. All the procedure was done under Ar atmosphere. The annealed samples were characterized by X-ray powder diffractometry using Cu K_α radiation (PANalytical X'Pert PRO) equipped with a fast linear detector. XRD data were collected in the 2θ° range 8-100° with a nominal scan rate of 120s step-1 and a step size of 0.016° at room temperature. Neutron diffraction data were collected at the Echidna high resolution diffractometer of The Brag institute (Sydney/Australia).[15]

Rietveld refinements of XRD powder patterns were performed starting from X-ray single crystal literature structure data by Deiseroth et al.[16] with the Generalized Structure Analysis System (GSAS) by Larson and von Dreele,[17] along with the graphical user interface EXPGUI[18].

Ionic conductivity measurements were carried out by impedance spectroscopy (Schlumberger Solartron SI1260) in the frequency range of 1 Hz to 10 MHz using stainless steel plate electrodes. At each temperature the samples were kept for 20 min for thermal equilibration. The bulk resistance R_b was determined from fitting impedance data to Nyquist plots. The equivalent circuit consists of C_b, R_b and a Warburg element.

$LiFePO_4$ was prepared using carbothermal reduction:[19] a stiochiometric amount of 1M LiH_2PO_4 (Aldrich , purity 99%) and 0.5M Fe_2O_3 (prepared from decomposition of Fe-oxalate) and 0.5 M carbon were ball-milled for 12h using a Spex Ball mixer (D8000) USA , then heated at 700°C for 4h in Argon, heating and cooling rate of 3 °C/min. The final powders were manually ground for further characterization.

RESULTS

Bond valence Lithium migration maps

Regions of low Lithium site energy, *E(Li),* in the structure models are expected to belong to pathways for Li^+ ion migration. *dc* conduction requires continuous pathways across the unit cell in at least one dimension. The migration pathways can be visualized as regions enclosed by isosurfaces of constant *E(Li)* based on calculations of *E(Li)* for a grid of hypothetical Li positions covering the entire unit cell with a resolution of *ca.* 0.1 Å. The threshold value of *E(Li)* for which isosurfaces form a continuous migration path (that includes both occupied and vacant Li sites), permits a rough estimate of the activation energy for Li^+ migration. As the approach neglects relaxation, the assessment of the activation energy is based on an empirical correlation observed for a wide range of Lithium ion conductors (cf. Figure 1).

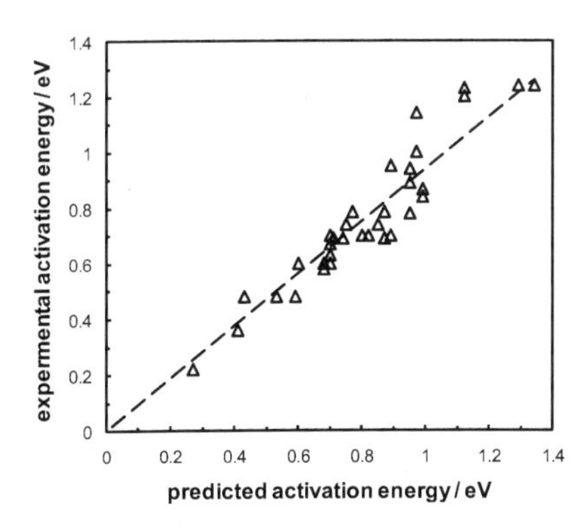

Figure 1. Empirical correlation between the migration energy predicted from the Li site energy thresholds along Li migration pathways and the experimentally observed activation energy.

Our analysis reveals significant differences to results of a recent geometric (Voronoi–Dirichlet partition based) study of cages and channels in crystalline Lithium oxides by Anurova *et al.*[21] Differences are particularly pronounced for the 33 types of ternary oxides listed in ref. 21 as containing 1D Li pathways: In our BV based models 1D migration channels with low to moderate activation energies are observed for 19 of these structures only, while 3 exhibit 2D pathways (LT-$LiPO_3$, $Li_2W_2O_7$, Li_2TeO_3), 6 even 3D pathways (α-Li_3BO_3, Li_4GeO_4, Li_2SeO_4, $Li_2T_2O_5$ (T=Si, Ge), Li_4TeO_5, $Li_4Mo_5O_{17}$) and in further 6 cases the structure models employed in ref. 20 are questionable, implausible or do not yield any paths. A main reasons for deviations is the complex curved nature of paths, which are difficult to identify from a geometric approach emphasizing straight channels. Unsurprisingly, a closer agreement is found for structures suggested to be 2D or 3D conductors in ref. 21. The main difference is however that the pathway analysis yields energy thresholds along the pathways and hence allows a direct estimate of activation energies.

Table 1 Average cation coordination numbers N_c in oxides, parameters for bond valence calculations (R_0, $b = 1/a$, cut-off distance R_{co}) and the resulting Morse potential parameters D_0, R_{min}, as determined from a wide range of stable oxides.

cation	N_c	$R_0/Å$	$b/Å$	$R_{co}/Å$	D_0 /eV	R_{min} /Å
H(1)	1.923	0.87045	0.457	4	1.88580	1.12768
Li(1)	5.021	1.17096	0.516	5.5	0.98816	1.94001
Be(2)	4	1.20903	0.541	5.5	2.76882	1.52217
B(3)	3.417	1.35761	0.385	4.5	2.38924	1.34003
C(4)	3	1.39826	0.447	5	4.79187	1.20089
C(2)	1	1.41368	0.415	5	2.40553	1.03098
N(5)	3	1.46267	0.45	5	6.27677	1.16142
N(3)	2	1.40795	0.448	5	3.81089	1.13758
NH4(1)	3.467	2.0338	0.442	6	0.40537	2.45364
Na(1)	6.52	1.56225	0.482	6	0.57523	2.37433
Mg(2)	5.897	1.48398	0.512	5.5	1.57554	1.95627
Al(3)	5.327	1.59901	0.424	5	1.80346	1.75806
Si(4)	4.1	1.60817	0.432	5	2.85720	1.53594
P(5)	4	1.62038	0.437	5	3.89635	1.44066
P(3)	3	1.51555	0.402	4.5	2.02062	1.41051
S(6)	4	1.6422	0.441	5	4.96726	1.38102
S(4)	3	1.64282	0.427	5.5	3.03672	1.41188
Cl(7)	4	1.67946	0.443	5	5.9910	1.34801
Cl(5)	3	1.69552	0.445	5.5	4.29089	1.35653
Cl(3)	2	1.72265	0.491	5.5	3.07119	1.38441
K(1)	8.846	1.94117	0.436	6	0.34985	2.76636
Ca(2)	7.544	1.79519	0.476	5.5	0.99429	2.32032
Co(2)	5.506	1.59773	0.451	5.5	1.51476	1.93362
Cu(3)	4	1.70964	0.427	5	1.88242	1.70823
Cu(2)	2.56	1.57422	0.449	5	1.85341	1.56633
Cu(1)	2.56	1.5873	0.341	5	0.66417	1.78269
Zn(2)	4.718	1.65344	0.403	5	1.24031	1.88557
Ga(3)	4.905	1.71606	0.373	5	1.18456	1.79391
Ge(4)	4.305	1.73939	0.396	5	1.91375	1.66872
As(5)	4.029	1.76689	0.411	5	2.71934	1.58127
As(3)	3	1.76706	0.404	5	1.51493	1.64554
Se(6)	4	1.79866	0.416	5.5	3.44865	1.53287
Se(4)	3	1.80095	0.427	5.5	2.38082	1.55957
Br(7)	4	1.83658	0.423	5.5	4.24339	1.50274
Rb(1)	10.02	2.08597	0.413	6.5	0.26813	2.89683
Sr(2)	9.4	1.95311	0.455	5.5	0.74351	2.53589
Y(3)	7.285	1.90384	0.478	5.5	1.62701	2.21523
Zr(4)	6.765	1.84505	0.49	5.5	2.19103	1.99602
Nb(5)	6.044	1.86588	0.498	5.5	2.72326	1.85459
Nb(4)	6	1.78543	0.526	6	2.7096	1.85989
Nb(3)	6	1.74581	0.501	6	2.02848	1.9519
Mo(6)	4.764	1.90934	0.391	5	1.9915	1.71254
Mo(5)	5.98	1.8476	0.482	5.5	2.64802	1.7867
Mo(4)	6	1.7239	0.562	6.5	3.10807	1.85099
Mo(3)	5.7	1.78933	0.418	5.5	1.42826	1.92974
Ru(6)	4.5	1.92579	0.425	5.5	2.42109	1.66431
Ru(5)	6	1.87442	0.436	5.5	2.13208	1.81571
Ru(4)	6	1.79363	0.449	5.5	1.99513	1.84053
Rh(4)	6	1.77675	0.403	5.5	1.62725	1.81793
Rh(3)	6	1.67013	0.478	5.5	1.92826	1.86915
Pd(4)	5.333	1.805	0.449	5.5	2.04218	1.79813
Pd(2)	4	1.62359	0.498	5.5	1.7391	1.83671
Ag(1)	4.438	1.78239	0.394	5	0.63519	2.22578
Cd(2)	6.176	1.83926	0.407	5.5	0.98346	2.1694
Ta(4)	5.5	1.75632	0.546	6	2.75655	1.79826
In(3)	6.024	1.90305	0.353	5	0.84076	2.02471
Sn(4)	6.069	1.89019	0.379	5	1.35268	1.93422
Sn(2)	3.325	1.87499	0.458	5.5	0.97261	1.9642
Sb(5)	6	1.89768	0.400	5.5	1.95523	1.86318
Sb(3)	6	1.92036	0.423	5	1.17786	2.07526
I(7)	5.8	1.92274	0.419	5.5	3.21424	1.74105
I(5)	3.1	1.97775	0.424	6	2.48947	1.64421
Te(6)	6	1.91343	0.412	5.5	2.56406	1.80876
Te(4)	3.396	1.9529	0.401	5.5	1.67169	1.75208
Cs(1)	11.79	2.25899	0.419	6.5	0.23307	3.13121
Ba(2)	10.32	2.15998	0.437	6	0.57994	2.73769
Sc(3)	6.255	1.7322	0.494	5.5	2.1561	1.99615
Ti(4)	6	1.72394	0.503	5.5	2.81333	1.83144
Ti(3)	6	1.69766	0.46	5.5	1.97851	1.88619
V(5)	4.166	1.79445	0.51	5.5	3.69533	1.60258
V(4)	5.738	1.74932	0.426	5	2.08047	1.77638
V(3)	6	1.67799	0.439	5.5	1.82936	1.85797
Cr(6)	4	1.82471	0.476	5.5	3.68751	1.53251
Cr(5)	4	1.76781	0.402	5.5	2.36551	1.55546
Cr(4)	5.429	1.76095	0.409	5.5	1.93329	1.76209
Cr(3)	6	1.66198	0.43	5.5	1.77335	1.83887
Mn(7)	4	1.87362	0.52	6.5	4.9163	1.48171
Mn(6)	4	1.82018	0.416	5.5	2.82236	1.52931
Mn(5)	4	1.78879	0.413	5.5	2.46456	1.57577
Mn(4)	5.923	1.73272	0.402	5	1.85886	1.77045
Mn(3)	5.862	1.68993	0.437	5.5	1.81283	1.85786
Mn(2)	5.91	1.62758	0.481	5.5	1.64143	2.02969
Fe(4)	6	1.76559	0.41	5.5	1.87285	1.82786
Fe(3)	5.733	1.7084	0.42	5	1.66681	1.86647
Fe(2)	5.743	1.57911	0.48	5.5	1.69269	1.96005
Ni(3)	6	1.64888	0.414	5.5	1.66191	1.81887
Ni(2)	5.933	1.5592	0.443	5.5	1.46841	1.92452
Co(3)	6	1.59234	0.434	5.5	1.87024	1.7762
La(3)	9.83	2.06392	0.451	5.5	1.18587	2.46989
Ce(4)	7.867	2.02821	0.443	5.5	1.48412	2.19872
Ce(3)	9.147	2.03118	0.449	5.5	1.22048	2.37861
Pr(3)	9.067	2.03652	0.439	5.5	1.17041	2.37113
Nd(3)	8.647	2.02425	0.428	5.5	1.13205	2.33016
Sm(3)	8.119	2.01168	0.433	5.5	1.17622	2.29536
Eu(3)	7.743	2.00469	0.434	5.5	1.19545	2.26888
Eu(2)	10.111	1.89158	0.494	6	1.13032	2.53846
Gd(3)	8.052	1.99654	0.415	5.5	1.09161	2.2719
Tb(4)	6	1.96244	0.494	6	1.70132	2.38506
Tb(3)	7.958	1.95675	0.433	5.5	1.20764	2.23563
Dy(3)	7.828	1.96029	0.426	5.5	1.1735	2.22689
Ho(3)	7.5	1.97099	0.415	5.5	1.12157	2.21122
Er(3)	7.135	1.95608	0.412	5.5	1.12394	2.17477
Tm(3)	6.912	1.94901	0.421	5.5	1.18138	2.16042
Yb(3)	6.875	1.92872	0.426	5.5	1.21989	2.1422
Lu(3)	6.83	1.91728	0.421	5.5	1.19488	2.136
Hf(4)	7.105	1.83361	0.478	6	1.89992	1.99964
Ta(5)	6.09	1.86816	0.486	5.5	2.36669	1.85532
W(6)	5.688	1.90641	0.401	5	1.84267	1.77713
W(5)	6	1.81975	0.498	6	2.6157	1.76261
W(4)	6	1.74558	0.520	6	2.47114	1.81945
Re(7)	4.098	1.97792	0.508	6	3.55593	1.59634
Re(6)	5.5	1.91007	0.498	6	2.95099	1.71147
Re(5)	6	1.82664	0.479	6	2.41099	1.76914
Re(3)	6	2.2071	0.401	6	0.81067	2.33218
Os(8)	5.333	1.97728	0.512	6	3.71019	1.66146
Os(7)	6	1.95775	0.479	5.5	2.91948	1.72869
Os(6)	6	1.93192	0.463	5.5	2.44871	1.7828
Os(4)	6	1.75302	0.498	6	2.27524	1.81244
Ir(5)	6	1.89791	0.479	6	2.32476	1.83476
Ir(4)	6	1.83233	0.436	5.5	1.68667	1.87402
Pt(4)	6	1.82198	0.479	5.5	2.03825	1.87174
Pt(2)	4	1.51205	0.574	5.5	2.14999	1.80179
Au(3)	4	1.81761	0.498	5.5	1.96967	1.81312
Au(1)	2	1.71819	0.441	5.5	0.85304	1.89543
Hg(2)	6.966	1.81276	0.465	5.5	1.12852	2.25275
Hg(1)	4.786	1.8128	0.465	5.5	0.73931	2.43155
Tl(3)	5.22	2.06297	0.338	5	0.67637	2.10642
Tl(1)	8.03	1.91752	0.483	6	0.34999	2.77086
Pb(4)	5.74	2.03293	0.354	5	1.02719	2.02857
Pb(2)	7.541	2.01825	0.433	5.5	0.63833	2.44191
Bi(5)	6	2.04498	0.371	5	1.4405	1.98599
Bi(3)	6.058	2.03677	0.414	5.5	0.97904	2.18321

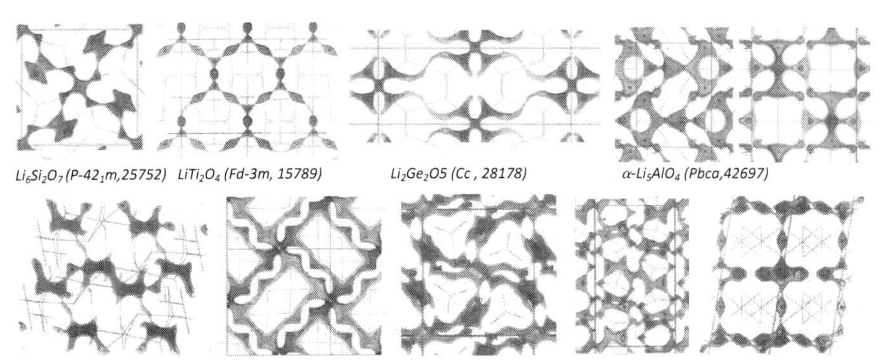

Figure 2 Examples of isosurfaces of constant *E(Li)* as bond valence models of Li^+ ion migration pathways in ternary oxides. In each graph three isosurfaces corresponding to increasing site energies are superimposed (red, magenta, light blue). Li atoms are indicated as blue crosses and labeled. Other atom positions are indicated by line or stick models. Labels below each graph indicate the respective compound name, space group and ICSD database[20] code of the underlying structure data.

MD simulations of $LiFePO_4$ and surface modified $LiFePO_4$ with the BV-based force-field

MD simulations have been conducted with GULP[22] using the Morse-type force field discussed above for models of bulk $Li_{0.99}FePO_4$ (with a single built-in $Li'_{Fe}/Fe^{\bullet}_{Li}$ antisite defect per 96 formula units), glassy $[(Li_2O)_{2/3}\ (P_2O_5)_{1/3}]_{411}$, for a 4367 atoms 3D-periodic interface structure $[LiFePO_4]_{320}$-$[(Li_2O)_{2/3}\ (P_2O_5)_{1/3}]_{491}$ for 300K ≤ T v< 1000 K and for a 5368 atoms model of $Li_{0.99}(Fe_{0.95}Na_{0.05})PO_4$.

To validate the bond-valence based force-field described above, we conducted temperature dependent NPT simulations in the range 300 – 1000 K and compared them to both our and literature experimental data (RT – 773K) on the thermal expansion of $LiFePO_4$ (see Figure 3). Our Rietveld refinements yield linear thermal expansion coefficients α_{11}= 1.22 ×10^{-5} K^{-1}, α_{22}= 1.47 ×10^{-5} K^{-1}, α_{33}= 1.89 ×10^{-5} K^{-1}). The resulting volume expansion coefficient β_{exp} (300 – 773 K) = 4.6 ×10^{-5} K^{-1} harmonizes with the simulation result β_{exp} (300 – 1000 K) = 5.3 ×10^{-5} K^{-1} for $LiFePO_4$. Simulations for the phase $Li_1Na_{0.05}Fe_{0.95}PO_4$ yield a 1% increase of the lattice constant and a volume expansion coefficient of 5.5 ×10^{-5} K^{-1}. Note also the nonlinear temperature dependence of the Fe-Fe distance (Fig. 3) that should affect the temperature dependence of the electronic conductivity.

For the Li diffusion analysis bulk and $LiFePO_4$: $Li_4P_2O_7$ heterostructure ensembles are equilibrated in NPT simulations over 600ps. The relaxed volume is then fixed and equilibration continued for 150ps at each temperature followed by 600 ps (for T=600 - 1000K), 1 ns (for T=500K), or 3ns (for T = 400 - 300K) production runs in 1.5 fs steps.[11,13] When constructing the interface model based on relaxed structure models of $LiFePO_4$ and glassy $Li_4P_2O_7$, the (010) surface of $LiFePO_4$ ⊥ to [010] Li^+ migration channels is chosen, as it is the most relevant surface for Li^+ transport and the most prominent face of $LiFePO_4$ nanoparticles. The extra charge in non-stoichiometric $Li_{0.99}FePO_4$ is distri-buted over all Fe cations emulating electron transfer on a time scale faster than that of ionic transport. Simulations for bulk Li_xFePO_4 (x = 95/96) yield diffusion coefficients, $D(Li)||b$, (cf. Fig. 4) that are consistent with values derived from conductivity data by Amin *et al.*,[23] but considerably higher than values reported by Li et al.[24] The simulated E_A = 0.57eV for bulk $Li_{.99}FePO_4$ is consistent with findings

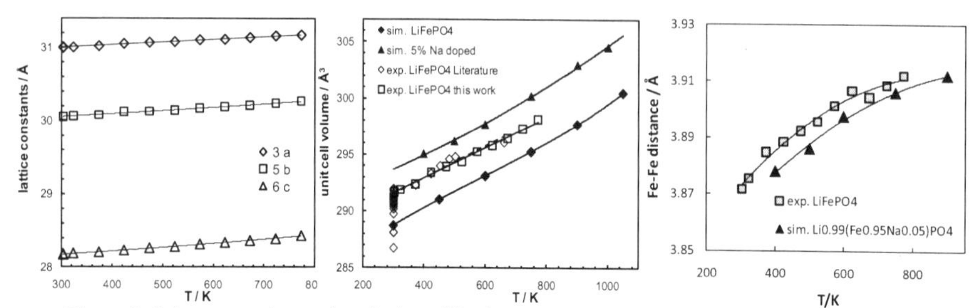

Figure 3: L.h.s.: experimental variation of lattice constants in $LiFePO_4$ from Rietveld refinements in the temperature range 303-773 K. Center: Variation of simulated (filled symbols) and experimental (open symbols) unit cell volume with temperature. R.h.s.: Temperature dependence of the Fe-Fe distance in pure and doped $LiFePO_4$ from Rietveld refinement and NPT simulations.

of various experimental and theoretical studies. A BV pathway analysis for both time-averaged and snapshot-type structure models of fully ordered $LiFePO_4$ harmonizes with ab initio studies in yielding zig-zag shaped one-dimensional Li^+ pathways $\|b$. The apparent contradiction to some of the experimental conductivity data implying a 2D nature of the Li^+ motion motivated our study on the effect of likely defect scenarios on the expected pathway dimensionality.

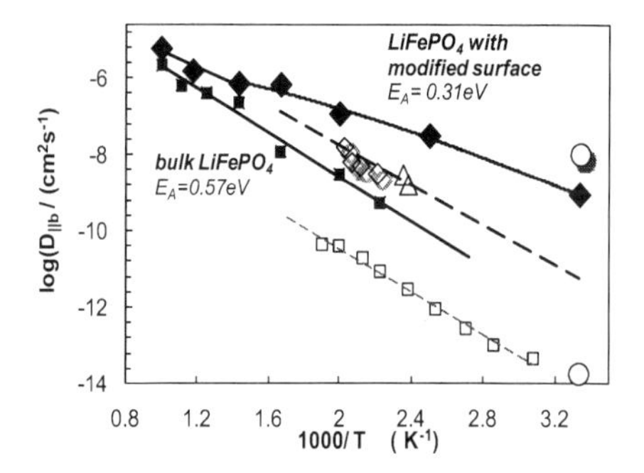

Figure 4 Comparison of simulation results for the Li^+ diffusion constant along the channel direction in LiFePO4 from this work (■: bulk $LiFePO_4$; ♦: surface-modified $LiFePO_4$) with literature data (open symbols, experimental data: ◊,Δ[23], □[24]).

Due to the moderate energetic disadvantage (≈1.1 eV)[25] some antisite defects will occur in equilibrated samples, but concentrations in real samples will be significantly higher due to non-equilibrium sample preparation routes and the vicinity of surfaces or interfaces. Antisite defects in $LiFePO_4$ have recently been visualized by scanning transmission electron microscopy.[26] As seen from BV pathways (cf. Fig. 5) for a model containing a single antisite defect pair, this defect affects local Li^+ paths in two ways: $Fe^{\bullet}_{Li}$ inside the channel block ion transport $\|$ b, while Li'_{Fe} opens up a new path $\perp b$ bridging Li channels. This is also seen in MD simulations for elevated temperatures (Fig. 5c, d). For sufficiently high antisite defect concentrations this may lead to long-range Li^+ transport perpendicular to the channels. Our Monte Carlo simulations suggest that the defect concentration required for a 2D percolating Li^+ pathway cluster can be reduced to about 2.2% if a significant energetic preference for the formation of antisite defect pairs close to existing defects is assumed[11] in line with the experimental TEM findings.

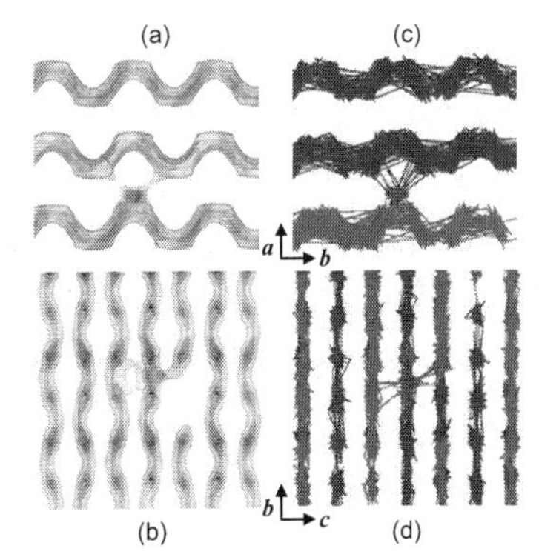

Figure 5 (a,b) Projections of Li^+ pathway models in $Li_{.99}FePO_4$ with 1 antisite defect. The three superimposed isosurfaces represent constant Li^+ site energy characteristic of equilibrium sites (black), migration channels || *b* (brown) and migration via antisite defects (beige). The BV based energy landscape approach yields low-energy paths even though at this temperature no transport $\perp b$ occurs during the simulated period of 1.5 ns. For comparison graphs (c, d) display Li^+ trajectories from MD simulations for T = 1000 K (1.5 ns). Atoms initially located in 1 b-c layer are marked in red.

Extraordinarily high (dis)charging rates were recently observed for cathodes with the nominal composition $LiFe_{0.9}P_{0.95}O_{4-\delta}$.[27] Electron microscopy showed that the ca. 50 nm thin nanocrystals are phase segregated into Li_xFePO_4 and a surface glass layer ($Li_4P_2O_7$ or a similar composition).[27-29] Phase segregation might also involve iron phosphides and/or Fe^{3+} in the glass, which would raise the electronic conductivity. Our MD simulations show that $D(Li)_{||b}$ in the $LiFePO_4$ layer is enhanced by ca. 3 orders of magnitude for typical working temperatures (Fig. 4). $D(Li)_{||b}$ in delithiated heterostructures is similar to the value for $x \approx 1$, while the anisotropy is less pronounced. Due to the reduced activation energy (= migration energy = 0.31eV for 300K $\leq T \leq$ 700K), the conductivity enhancement practically vanishes for T > 700K, where E_A approaches the bulk value. Li^+ ions are significantly enriched on the Li_xFePO_4-side of the interface and depleted on the $Li_4P_2O_7$ glass side (Fig. 6). The overall Li^+ concentration in the Li_xFePO_4 layer increases by 4-7 % ("x=1") or 17 - 37% ("x=0.06") - a significant deviation from local electroneutrality within each phase.

The change in the extent of Li^+ redistribution with x implies a fast pseudo-capacitive energy storage component. Moreover, antisite formation near the interface boosts Li^+ mobility $\perp b$ and promotes a full utilization of the (dis)charging capacity (but slows down transport || *b*). A layer-by-layer analysis reveals that Li^+ mobility $||b$ is enhanced in the interface region (compared to bulk values) by a factor of 60 (T=600K) to ca. 3000 (T=300K) in harmony with the maximum of the volume fraction of low BV mismatch regions at the interface. The relative enhancement is highest in Li_xFePO_4, while for T $\geq$ 600 K $D(Li)_{||b}$ is higher in the glass phase. The anisotropy $D(Li)_{||b} / D(Li)_{\perp b}$ is gradually reduced from practically infinite in $LiFePO_4$ to (trivially) one within the glass (partly via antisite defects, partly via Li^+ crossing the interface). Over ca. 1nm from the interface $D(Li)_{\perp b}$ in $LiFePO_4$ drops by a factor $\approx$90. As for "x = 0.06" most Li^+ reside close to the interface, the overall anisotropy of *D(Li)* becomes a function of x.

The effect of homogeneous doping on Fe sites has been explored by MD simulations for the system $Li_{0.99}(Fe_{0.95}Na_{0.95})PO_4$ (5368 atoms, NVT simulations with relaxed lattice constants at each temperature in the range 500 – 900 K, simulated period 0.5 – 2 ns). As shown in Fig. 7 it was found that the mobility of the Na^+ ions that had originally been placed on Fe^{2+} sites significantly promotes the mobility of Li^+ ions perpendicular to the channels (which is too low to be observed in the Na-free model) and moreover slightly reduces the activation energy along the channel direction to 0.41 eV (instead of 0.57 eV). The effect may be understood in analogy to the $Li_{Fe}^{/}$ antisite defects: $Na_{Fe}^{/}$ is sufficiently mobile to allow for Li^+ transport perpendicular to the channels and at least at elevated temperatures is also mobile along the channels so that it does not block them in the same way as $Fe_{Li}^{\bullet}$ would do. For the highest simulated temperature (900 K) Na^+ has effectively nearly the same mobility

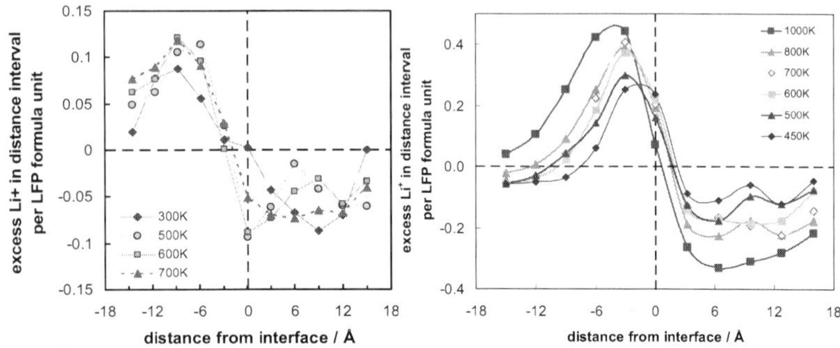

Figure 6. Li^+ redistribution at the Li_xFePO_4:$Li_4P_2O_7$ interface displayed as % excess Li^+ in ca. 3Å thick slices vs. distance of the slice from the interface for different temperatures (l.h.s.: x=1; r.h.s.: x = 0.06). Li^+ ions are enriched in the Li_xFePO_4 phase (negative distances), depleted in the $Li_4P_2O_7$ glass. Each value represents an average over 100 - 250 MD time steps.

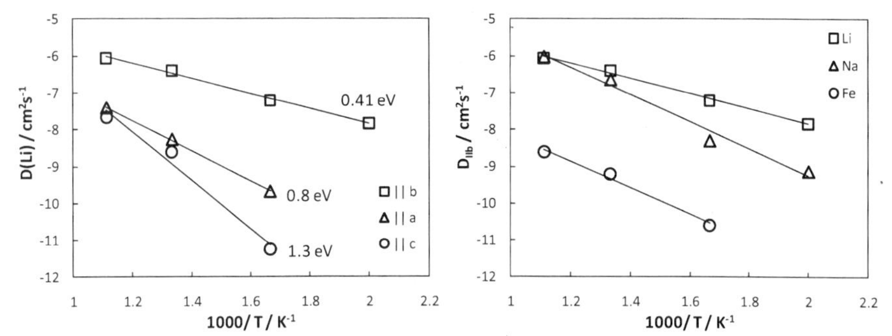

Figure 7: l.h.s.: Simulated diffusion coefficients for Li^+ cations in $Li_{0.99}(Fe_{0.95}Na_{0.95})PO_4$ along the crystallographic axes a – c. The highest diffusion coefficients occurs as expected along the channel direction (|| b) with a slightly reduced activation energy of 0.41 eV when compared to $Li_{0.99}FePO_4$, and in contrast to the latter the Na

as Li^+ (preferably migrates along the channels), while for lower temperatures the mobility of Na^+ exhibits a more pronounced decrease (activation energy || b 0.7eV) and becomes nearly isotropic, so that the total 3D mobility of Li^+ and Na^+ remains similar. A minor local mobility of Fe^{2+} can only be observed at the highest temperatures.

Pathways in high energy LIB cathode materials $LiVPO_4F$ and $LiFeSO_4F$

Among the framework materials based on phosphate or sulfate polyanion building blocks that are increasingly regarded as favourable replacements for conventional oxide-based cathode materials, lithium vanadium fluorophosphates, $LiVPO_4F$ has been investigated recently by various groups.[30-32] $LiVPO_4F$ is iso-structural with the naturally occurring triclinic minerals tavorite ($LiFePO_4OH$) and amblygonite ($LiAlPO_4F$). The $LiVPO_4F$ structure comprises a 3-D framework built up from single chains of 2 distinct corner-sharing [VO_4F_2] octahedra cross-linked via [PO_4] tetrahedra[30] wherein O atoms are shared between both environments. The strong inductive effect of the PO_4^{3-} anion, combined with the presence of structural F^-, permits the reversible lithium insertion reactions for $Li_{1-x}VPO_4F$ to occur at the unusually high operating potential of ca. 4.2V versus Li. More recently, analogous fluoro-sulfates $LiMSO_4F$ (M=Fe, Co, Ni) have also been synthesized[33-36] and found to be isostructural to $LiMgSO_4F$ and $LiVPO_4F$ (though a different setting of the unit cell is chosen in the original publication). As $LiMSO_4F$ (M = Fe, Co, Ni etc.) is unstable at the temperatures of most solid state reactions ($LiFeSO_4F$ decomposes for T $\geq$ 350 C),[33] it became accessible only via ionothermal synthesis in EMI-TFSI or more recently in tetraethylene glycol.[36] Cells with $LiFeSO_4F$ cathodes and Li anodes deliver 85% of the theoretical specific capacity of 151 mAh g^{-1} at C1 rate without carbon coating or nanostructuring. *ac* conductivity of $LiFeSO_4F$ was found to be mainly ionic with $E_A = 0.99 \pm 0.01$ eV and $\sigma(300K) \approx 10^{-10}$ S cm^{-1}. From projections of the framework structure, Recham *et al.*[33] originally suggested 3 tunnels with large cross sections (along [100], [010], and [101]) as pathways for a presumed 3D Li^+ migration.

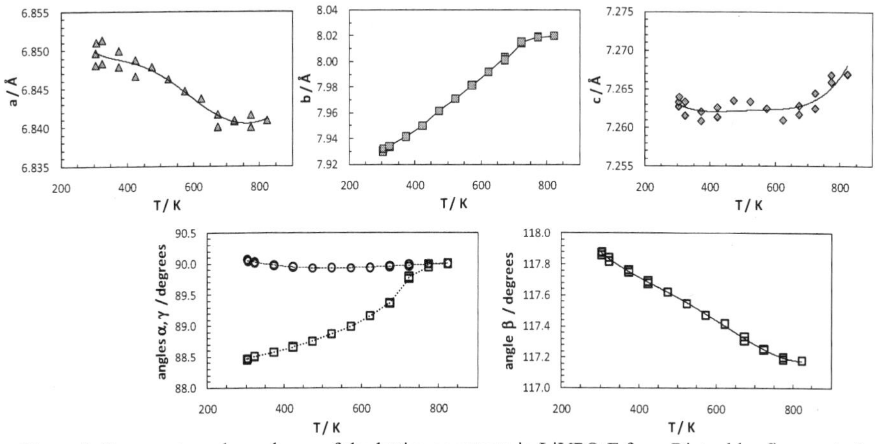

Figure 8: Temperature dependence of the lattice constants in $LiVPO_4F$ from Rietveld refinement of powder XRD data revealing the triclinic to monoclinic phase transition. To emphasize the structural relationship between the phases a C-centered setting of the triclinic unit cell has been chosen.

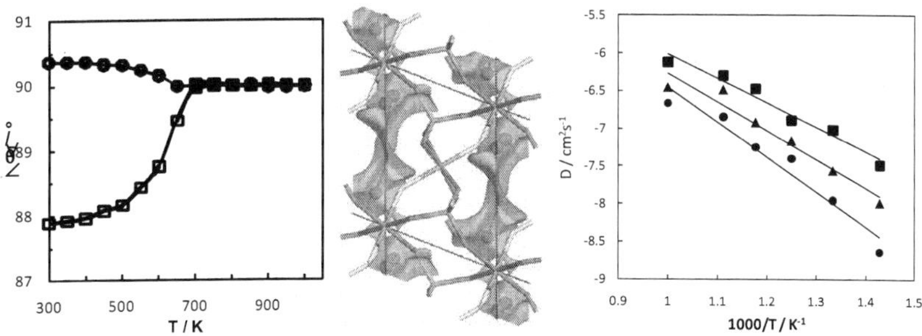

Figure 9: L.h.s.: temperature dependence of the triclinic angles α (o) and γ ( ) in $LiVPO_4F$ from NPT Molecular Dynamics simulations. Center: BV model of Li pathways of lowest activation energy in $LiVPO_4F$. R.h.s.: Simulated Li diffusion coefficients in (the high temperature phase of) $LiVPO_4$ along the fast axis (||c; ■) and the two perpendicular directions (●, ▲).

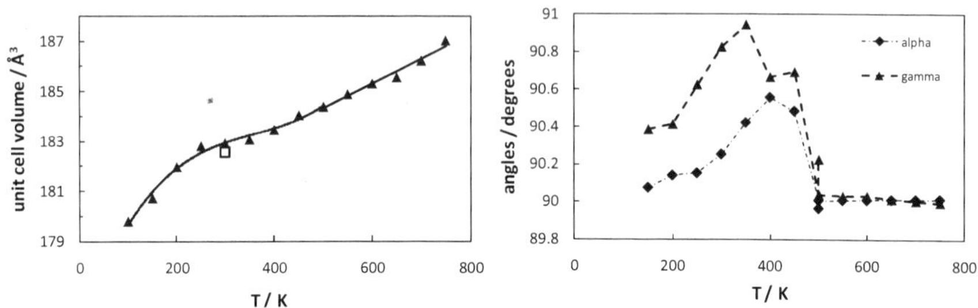

Figure 10: Temperature dependence of the primitive unit cell volume (l.h.s.) and of the triclinic angles α and γ for a pseudo-orthorhmbic supercell (r.h.s.) of $LiFeSO_4F$ from NPT Molecular Dynamics simulations. Full symbols indicate NPT MD simulation results, while the open square indicates the experimental unit cell volume at room temperature.

Our Rietveld refinements of $LiVPO_4F$ in the temperature range revealed that $LiVPO_4F$ undergoes a second order phase transition at about T = 750 K from a triclinic phase to a monoclinic phase (C2/c). To emphasize the structural relationship between both phases the non-standard space group C-1 has been chosen here (as commonly practice in the literature for transitions of related phases).

MD simulations of a 2048 atom supercell of the C-1 unit cell accordingly yield a monoclinic to triclinic phase transition, yet with a slightly lower transition temperature (ca. 700 K). The diffusion coefficient in the monoclinic high temperature phase is highest along channels extending along the z direction ($E_A \approx 0.4$ eV), while the perpendicular directions exhibit about 0.75 eV or 0.9 eV and for T= 700 K about 0.5 or 1 order of magnitude lower diffusion coefficients.

A similar structural phase transitions is also observed in $LiFeSO_4F$, as seen from the change in the thermal expansion (Fig. 10). Again it can be described as a triclinic to monoclinic phase transition.

An average cell based on the relaxed MD simulated supercell at 300 K a (Fig. 11) closely resembles the published XRD data. Besides a minor rotation of SO_4^{2-} the main difference is in the Li^+ distribution, which is characterized by a pronounced disorder along channels extending along [111] with two weak Li density maxima: Li(1) at 0.145, 0.571, 0.633 (occupancy n = 0.6); and Li(2) at 0.450, 0.854, 0.854 (n = 0.4).

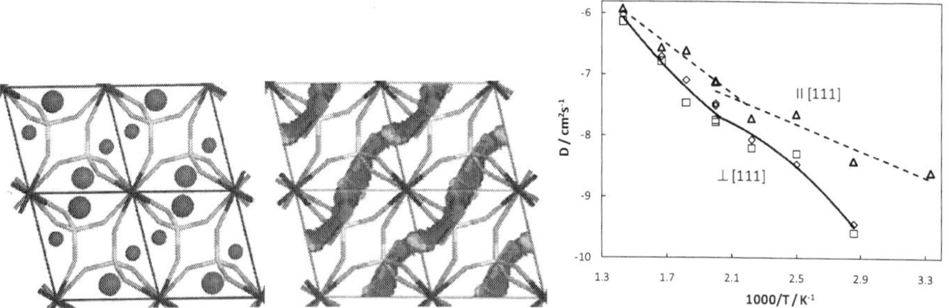

Figure 11. L.h.s.: MD structure model of $LiFeSO_4F$ projected on *ab* plane (Fe(1) green, Fe(2) blue; S yellow; F light blue, O orange, Li red (occupancy 0.6 for Li(1) = large spheres, 0.4 for Li(2) = small spheres). Center: Comparison of isosurface of constant Li density (dark) from MD trajectory (T = 600K) and BV pathway model (light) in the same projection. R.h.s.: Anisotropy of the diffusion coefficient in $LiFeSO_4F$ as a function of temperature.

These sites form Li(2)-Li(1)-Li(1)-Li(2) channels || [111] with distances ranging from 1.92 – 2.12 Å favourable for Li^+ transport, while migration in other directions requires hop distances ≥3.69Å. Static BV models for the Rietveld-refined and our MD simulated structure models [13] accordingly suggest zig-zag shaped 1D paths || [111] involving both Li sites as pathways of lowest migration energy barrier (≈ 0.22 eV, Fig. 3), while a migration energy of ca. 0.97eV would be required to connect the channels in the [010] direction and an only slightly higher activation energy of 1.1eV leads to a 3D network of Li^+ paths. Since low energy pathways || [111] connect partially occupied Li sites, a defect formation should not be required for migration along these channels.

NPT MD simulations again point towards a change in activation energy related to the monoclinic to triclinic phase transition. For the triclinic ambient temperature phase a migration energy of only 0.21 eV for the transport along the [111] channels is found in perfect accordance to the BV prediction (see r.h.s. graph in Fig. 11,). The experimental E_A = 0.99 eV is much higher than the one predicted for migration along [111] channels, but closely resembles the E_A for the formation of a 2D pathway network (paths along [111] and [010]).

Li Argyrodites as high power thiophosphate solid electrolytes

XRD patterns of samples that have been ball milled for more than 24 hrs show only partial crystallinity with broad peaks referring to the reactand phases P_2S_5 and LiX (X = Cl, Br, I) and sharper peaks for Li_2S. After annealing all samples yield the respective argyrodite phases with comparable crystallinity. Thus when compared to the original recipe that takes 7 days,[16] preparation time for argyrodites could be drastically reduced. Figure 12 shows a simultaneous Rietveld fit to XRD and neutron data[15] for the final Li_6PS_5I system as an example.

All the annealed samples of Li_6PS_5Cl, Li_6PS_5Br and Li_6PS_5I exhibit the high symmetry aristotype of the argyrodite structure (cf. Fig. 13, space group F-43m). The lattice parameters of Li_6PS_5Cl, Li_6PS_5Br and Li_6PS_5I are 9.856(4) Å, 9.983(6) Å and 10.145(3) Å, respectively in harmony to previously reported values. The linear expansion coefficients decrease from 3.4×10^{-5} K^{-1} for X=Cl via 3.1×10^{-5} K^{-1} (X=Br) to 2.5×10^{-5} K^{-1} for X=I (cf. Fig. 12). An apparent decrease of the P-S bond distance with increasing temperature (for X=Br,I) points towards a significant libration of the PS_4^{3-} ions in this structure at elevated temperatures. For X=Br[16] and Cl[15] sulfide and halide ions exhibit a mixed occupancy for two of the three anion sites, while the anion directly bonded to P on the third anion site is just sulfide.

Temperature dependent ionic conductivity of all samples as determined from impedance spectra showed essentially Arrhenius nature. Before annealing, ball milled Li6PS5X samples have exhibit ionic conductivity 3.3×10^{-6} (X=Cl), 3.2×10^{-5} (X=Br) and 2.2×10^{-4} S/cm (X=I) with activation energies 0.38, 0.32 and 0.26 eV respectively. The room temperature ionic conductivities of final crystalline compounds are 7.5×10^{-4} with E_A= 0.11eV (X=Cl), 7.3×10^{-4} with E_A=0.17eV (X=Br), and 4.6×10^{-7} S/cm with E_A = 0.25eV (X=I) respectively (see Fig. 13).

Li^+ ion migration pathways in the Li_6PS_5X argyrodite structures with X=Cl, Br and I are shown in Fig. 14. Roughly speaking, the three-dimensional pathway network for long range (dc) ion conduction in all Li_6PS_5X phases consist of interconnected low-energy local pathway cages. In detail however these cages and the way in which they are interconnected differs for the compounds with different halide ions. In the anion-disordered structures with X= Cl and Br (l.h.s. column in Fig 14) the sites of lowest energy agree with the experimentally refined half-occupied Li1 positions. The lowest energy short range pathway then interconnects 3 such partially occupied Li sites via 3 interstitial sites forming a pathway hexagon (E_A=0.18eV). Four such hexagons are then interconnected at a slightly higher energy threshold via a second interstitial site to form an extended pathway cage around the nominal Cl position (E_A=0.22eV) and the 3D long range pathway network is finally established by a direct connection in-between the cages (E_A=0.35eV). For the case of Li_6PS_5I however, experimental studies suggest a more disordered Li distribution described in ref. 16 by Li1 - Li2 - Li1 triplet sites. In our BV

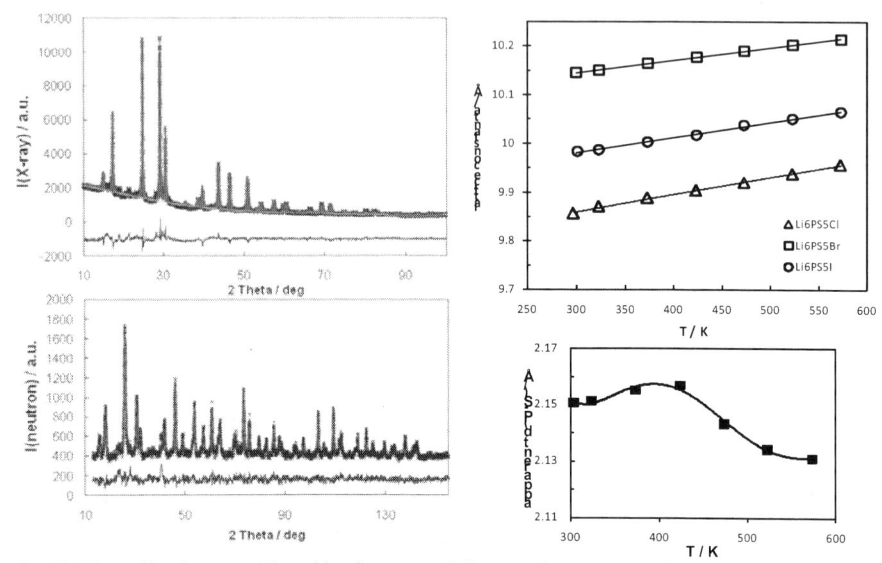

Fig. 12 L.h.s.: Simultaneous Rietveld refinement of X-ray and neutron powder diffraction data of the structure of Li_6PS_5I.[15] Upper r.h.s.: thermal expansion of Li_6PS_5X phases (X = Cl, Br, I) from in-situ XRD powder data. Lower r.h.s.: Apparent variation of P-S bond distance with temperature from powder XRD data for X=I.

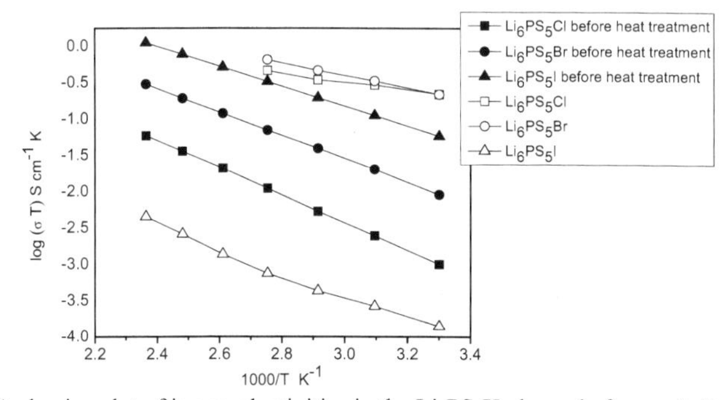

Figure 13. Arrhenius plot of ion conductivities in the Li_6PS_5X phases before and after annealing.

model the sites of lowest E(Li) accordingly imply a disordered Li distribution over the Li1 sites. The lowest energy interconnection between sites then includes the Li2 site in-between pairs of Li1 sites (0.09eV), but the Li2 position appears not to be a local minimum of E(Li), which is also in line with a literature difference Fourier plot. Experimental NMR data find an even lower activation energy of 0.043 eV for unspecified local hops.[16] Six of these dumbbell-like local pathways (which correspond to the intercage pathways in Li_6PS_5I) then form a pathway cage around the S2 site (not as in the previous case around the halide ion) with E_A=0.15eV. The cages are finally interconnected via an interstitial site for E_A=0.33eV in remarkable agreement with the experimental value 0.32 eV from low temperature impedance spectroscopy and 0.30eV from MD simulations. MD simulations also suggest 0.14eV for a localised motion.[37]

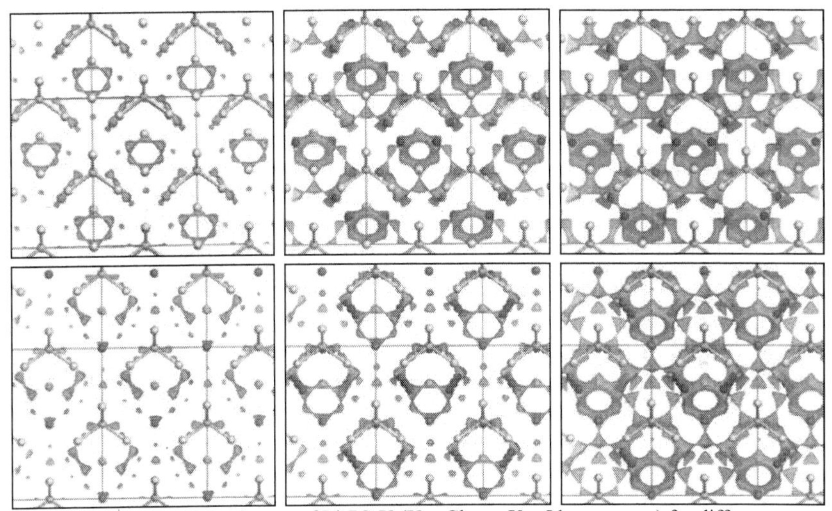

Figure 14. Li^+ migration pathways of Li_6PS_5X (X = Cl top, X = I bottom row) for different energy thresholds characterising, low-energy local Li^+ paths (l.h.s.), extended local Li^+ path cages (center), and paths for long range Li^+ migration (r.h.s.). Note that pathways are different in detail (cf. text).

CONCLUSIONS

The bond valence approach has been reworked into an effective local force-field that can be used both to analyse ion transport pathways and for MD simulations. $LiVPO_4F$ and $LiFeSO_4F$ are found to be quasi-one dimensional Li^+ ionic conductors (along channels that for $LiFeSO_4F$ extend along [111]). The experimental activation energy and power performance is however controlled by the moderate activation energy for transport perpendicular to the low energy pathways, which (as typical for 1D channels) will in most cases be blocked by defects, except for nanostructured materials.

The approach also helps to understand the effect of homogeneous and heterogeneous defects on the ionic transport in $LiFePO_4$, highlighting that the dimensionality of ionic motion will depend on the concentration and spatial distribution of antisite defects. Surface modification of Li_xFePO_4 nanocrystals by glassy $Li_4P_2O_7$ causes a significant Li^+ redistribution entailing an enhancement of Li^+ mobility and an thus an enhancement of room temperature Li^+ ion diffusion by about 3 orders of magnitude along the channels. This and the drastically enhanced mobility perpendicular to Li channels will facilitate a fast (dis)charging of Li batteries based on surface-modified $LiFePO_4$ electrodes. The pronounced change in the extent of the Li^+ redistribution from the phosphate glass layer into subsurface Li_xFePO_4 with x constitutes an ultrafast pseudocapacitive storage component.

Li_6PS_5X (X=Cl, Br) argyrodite structures were prepared by mechanical milling followed by annealing of the samples. This allows for a faster synthesis than the earlier reported long annealing. The obtained crystalline phase showed ionic conductivity of the order of 10^{-3} S/cm at ambient temperature for argyrodites exhibiting anion disorder (X=Cl, Br). Bond valence based Li^+ ion migration pathways for both compounds yield pathways with similarly low activation energies for long range transport, despite the differences in the detailed pathway topology for different halide ions. The extremely low activation energies reported from NMR studies refer to hops within extended local pathway cages that are not directly relevant for dc conductivity. Due to the high ionic conductivity of the compounds at room temperature, they are of high technological interest as one of the best solid electrolytes for all solid state battery applications.

ACKNOWLEDGEMENTS

Financial support in the frame of the Singapore Ministry of Education Grant MOE2009-T2-1-065 is gratefully acknowledged.

REFERENCES

[1] S. Adams, *J. Power Sources*, **159**, 200 (2006).
[2] J. Swenson and S. Adams, *Phys. Rev. B*, **63**, 054201 (2000).
[3] C. Müller E. Zienicke, S. Adams, J. Habasaki and P. Maass, *Phys. Rev. B*, **75**, 014203 (2007).
[4] S. Adams and R. Prasada Rao, *Phys. Chem. Chem. Phys.*, **11**, 3210 (2009).
[5] I.D. Brown, *Chem. Rev.* **109,** 6858 (2009).
[6] S. Adams and E. S. Tan, *Solid State Ionics*, **179**, 33 (2008).
[7] S. Adams, *Solid State Ionics*, **177**, 1625 (2006).
[8] S. Adams, *Acta Crystallogr. B, Struct. Sci.* **57**, 278 (2001).
[9] S. Adams, *softBV* ver. 0.96, 2004, http://www.softBV.net.
[10] I. D. Brown, *Acta Crystallogr. B, Struct. Sci.* **48**, 553 (1992).
[11] S. Adams; *J. Solid State Electrochem.* **14**, 1787 (2010).
[12] R. Prasada Rao, T.D Tho and S. Adams, *Solid State Ionics* **181**, 1–6 (2010).
[13] S. Adams and R. Prasada Rao; *Solid State Ionics*, *in press*.
[14] J. Barker, M.Y. Saidi, J. Swoyer, *J. Electrochem.Soc.,***150** A1394 (2003).
[15] R. Prasada Rao, N. Sharma, V. Peterson, M.V. Reddy, S. Adams; *in preparation*.
[16] H.-J. Deiseroth, S.-T. Kong, H. Eckert et al. *Angew. Chem.*, **120**, 767(2008).
[17] A.C. Larson, R. B. von Dreele, General Structure Analysis System (GSAS); Report LAUR 86-748; Los Alamos National Laboratory: Los Alamos, NM, 2000.
[18] B.J. Toby, *J. Appl. Crystallogr.* **34**, 210 (2001).
[19] H. Liu et al., *J. Solid State Electrochem.* **12**, 1011 (2008).
[20] ICSD Inorganic Crystal Structure Database, FIZ Karlsruhe.
[21] N.A. Anurova et al., *Solid State Ionics*, **179**, 2248 (2008).
[22] J.D. Gale, J. Chem. Soc. Faraday Trans. **93,** 629 (1997) .
[23] R. Amin, J. Maier et al., Lin, *Solid State Ionics*, **179**, 27 (2008) and **178,** 1831 (2008).
[24] J. Li et al., *Solid State Ionics*, **179**, 2016 (2008).
[25] C.A.J. Fisher, V.M. Hart Prieto and MS. Islam, *Chem. Mater.* **20**, 5907 (2008).
[26] S.Y. Chung et al. *Phys. Rev. Lett.* **100**, 125502 (2008).
[27] B. Kang, G. Ceder, *Nature* **458**, 190 (2009).
[28] G. Ceder and B. Kang, *J. Power Sources*, **194**, 1024 (2009).
[29] K. Zaghib et al. *J. Power Sources* **194**, 1021 (2009).
[30] J. Barker, R. K. B. Gover, P. Burns et al., *J. Electrochem. Soc.*, **152,** A1776 (2005).
[31] J. Barker et al., *J. Power Sources* **174**, 927 (2007).
[32] S.-C. Yin, P. Subramanya Herle, A. Higgins et al., *Chem. Mater.* **18**,1745 (2006).
[33] N. Recham et al., *Nature Materials*, **9**, 68 (2010).
[34] J-M. Tarascon et al., *Chem. Mater.*, **22**, 724 (2010).
[35] P. Barpanda et al.; *J. Mater. Chem.*, **20**, 1659 (2010).
[36] R. Tripathi et al., *Angew. Chem. Int. Ed.* **49**, 8738 –8742 (2010).
[37] O. Pecher, S.T. Kong, T. Goebel, et al., *Chem. Eur. J.*, **16**, 8347 (2010).

LITHIUM ION CONDUCTIVE SOLID ELECTROLYTE WITH POROUS / DENSE BI-LAYER STRUCTURE FOR ALL SOLID STATE BATTERY

Kiyoshi Kanamura, Ryo Oosone, Hirokazu Munakata, and Masashi Kotobuki

Department of Applied Chemistry, Graduate School of Urban Environmental Sciences
Tokyo Metropolitan University
Minamiosawa 1-1, Hachioji, Japan

ABSTRACT

Fabrication of all-solid-state Li battery with inflammable ceramics electrolyte has been strongly required to overcome safety issue of present Li battery. One of promising structures for ceramics electrolyte in all-solid-state battery is bi-layered structure composed of 3 dimensionally ordered macroporous (3DOM) layer and dense layer. In this study, we prepared $Li_{0.35}La_{0.55}TiO_3$ (LLT) ceramics electrolyte with the bi-layered structure by suspension filtration method. Thicknesses of the dense and the porous layers were about 23 and 105 μm, respectively. The porous layer involved uniform pores of 1.8 μm in diameter. An electrochemical property of $LiMn_2O_4$ / bi-layered LLT composite prepared by solvent substitution method was tested. A discharge and charge behavior of the composite electrode was clearly observed. From this result, it can be said that the composite electrode can be applied to rechargeable battery. The discharge capacity of the composite was 27 mA h g^{-1}.

INTRODUCTION

Present rechargeable Li ion batteries using liquid electrolyte suffer from severe safety concerns.[1] The most serious problem is flammability of the liquid electrolyte.[2,3] The all-solid-state rechargeable Li battery with ceramics electrolyte has been recognized as one of alternative technologies to overcome the safety issue.

However, one of the problems for all solid-state lithium-ion battery using ceramic electrolyte is high internal resistance because of a poor contact between solid electrolyte and solid active material[4]. We have proposed a novel electrode system using ceramic electrolyte with three-dimensionally ordered macroporous (3DOM) structure having a large surface area[5]. By injection of active material into the 3DOM pore, large contact area of active material / ceramics electrolyte can be obtained. In our group, preparation of ceramics electrolyte with 3DOM structure and development of an efficient method to inject an active material into the 3DOM pore have been studied[3-5]. $Li_{0.35}La_{0.55}TiO_3$ (LLT) and $Li_{1.5}Al_{0.5}Ti_{1.5}(PO_4)_3$ (LATP) ceramics electrolytes with 3DOM structure have been prepared by the suspension filtration method and colloid crystal template method, respectively[6,7]. Additionally, new impregnation procedure of active material into the 3DOM pores so-called "solvent substitution method"[7] has been developed. In fact, $LiMn_2O_4$ / 3DOM LLT electrode prepared by the solvent substitution method showed relatively high discharge capacity, 83 mA h g^{-1} that is 56 % of theoretical capacity.

By the way, a new family of Li conductive ceramics with a garnet-type structure has been reported by Weppner et. al[8,9]. Among them, $Li_7La_3Zr_2O_{12}$ (LLZ)[10] and $Li_5La_3Ta_2O_{12}$ (LLTa)[8] has been much attention because of their stablity against molten Li metal, implying that the Li metal anode can be used to LLZ and LLTa. One of the promising structures of the all-solid-state battery with the Li metal anode is supposed to be bi-layered structure composed of 3DOM porous and dense layers (fig. 1). The 3DOM porous layer with active material works as a composite electrode, and the dense layer can be utilized as an electrolyte. This bi-layered configuration makes us to fabricate the all-solid-state battery easily. In this case, only impregnation of cathode material into the pores and set of Li metal bottom of the dense layer are needed.

In this study, a fabrication of bi-layred LLT as a framework of the all-solid-state battery with Li metal anode was performed. Although LLT was not suitable electrolyte for Li metal anode because of facile reduction of Ti^{4+} in contact with Li metal anode[11], the LLT electrolyte was selected in this study based on our previous study on 3DOM electrolyte preparation. Here, an electrochemical property of $LiMn_2O_4$ / bi-layered LLT was also reported.

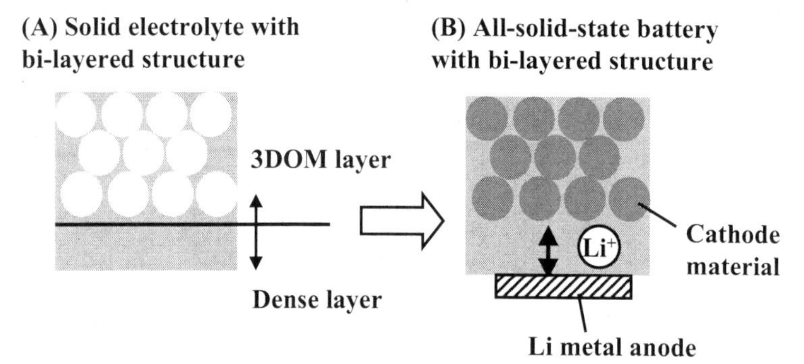

Fig. 1 Illustration of all-solid-state battery with bi-layered structure.

EXPERIMENTAL

$Li_{0.35}La_{0.55}TiO_3$ (LLT) solid electrolyte powder for fabrication of the bi-layered LLT was prepared by sol-gel method according to our previous paper[7]. The Li-La-Ti-O sol as a precursor of LLT was prepared from CH_3COOLi, $La(CH_3COO)_3 \cdot 1.5H_2O$, $Ti(OCH(CH_3)_2)_4$, CH_3COOH, i-C_3H_7OH, and H_2O in molar ration of 0.35 : 0.55 : 1 : 10 : 20 : 140. The sol was dried at 80 °C overnight and then calcined at 450 °C for 1 h to obtain amorphous LLT powder.

The bi-layered LLT was prepared by the suspension filtration method[6]. For preparation of porous layer, an ethanol suspension composed of LLT powder and monodispersed polystyrene (PS) beads with

3 μm in diameter (MAGSPHERE Inc.) was filtered under a small pressure difference of 5 kPa using a PTFE filter with 18 mm in diameter and pore size of 100 nm (ADVANTEC). Then, LLT powder suspended in ethanol was filtered to prepare the dense layer. Obtained composite membrane was calcined at 450°C for 1h to remove PS template and then 1100°C for 10 h. Morphology of bi-layered LLT was observed by a mean of scanning electron microscope (SEM, JEOL). Crystal structure was identified with X-ray diffraction (XRD, RINT-Ultima, Rigaku). Lithium ion conductivity of the bi-layered LLT was measured by the AC impedance measurement. Ag paste was painted on both sides of the bi-layered LLT to ensure electrical contact. The AC impedance measurement was performed at 5 mV voltage signal in a frequency range of 5–1 MHz. The Li ion conductivity of the bi-layered LLT was estimated by an equation shown in below

$$\frac{1}{R_t} = \sigma \times \frac{1 - \rho}{1 + \rho}$$

Here, σ is Li ion conductivity of bi-lyered LLT and R_t and ρ mean resistance and porosity, respectively.

Impregnation of $LiMn_2O_4$ into pores of the porous layer in the bi-layered LLT was performed by the solvent substitution method using sodium dodecyl sulfate (SDS) solution and Li-Mn-O sol[7]. The Li-Mn-O sol as a precursor of $LiMn_2O_4$ was prepared from CH_3COOLi, $Mn(CH_3COO)_2 \cdot 4H_2O$, $i\text{-}C_3H_7OH$, CH_3COOH, and H_2O (molar ratio = 1.1 : 2 : 20 : 40 : 70)[12]. The pores in the porous layer were filled with 0.02 wt% SDS solution. The bi-layered LLT filled with the SDS solution was immersed in the precursor sol for 1 h and dried at 70 °C for 10 min followed by calcination at 450 °C for 1 h. This procedure was repeated 4 times and finally the bi-layered LLT with Li-Mn-O was calcined at 700 °C for 1 h to obtain $LiMn_2O_4$ / bi-layered LLT composite.

In order to evaluate electrochemical property of the $LiMn_2O_4$ / bi-layered LLT composite, a half cell of $LiMn_2O_4$ / bi-layered LLT using Li metal anode was constructed. To avoid enhancement of electronic conductivity of LLT electrolyte, polymethylmethacrylate (PMMA) gel electrolyte was set between bottom of dense layer and Li metal anode as a buffer layer. In future, this layer should be changed to ceramic solid electrolyte layer, such as LLT and LLTa. The PMMA gel was prepared by polymerization of a mixture of methylmethacrylate monomer, ethylene glycol dimethacrylate, azobisisobutyronitrile, and 1 mol dm^{-3} $LiClO_4$ (in ethylene carbonate : diethyl carbonate = 1:1 vol.%) with the weight ratio = 1 : 0.05 : 0.02 : 2.87. Thickness of the PMMA gel was 300 μm. The electrochemical performance of the half cells was evaluated by the galvanostatic charge/ discharge test. Cut off voltage were 3.0 and 4.3 V vs. Li / Li^+ for discharge and charge, respectively, and all charge and discharge rates were 0.05 C.

RESULTS AND DISCUSSION

Figure 2 shows cross-sectional SEM images of the bi-layered LLT. The bi-layered LLT with dense and porous layers was fabricated successfully by the suspension filtration method. Thicknesses of the dense and the porous layers were about 23 and 105 μm, respectively. These thicknesses were consistent with ones estimated from amounts of filtrated LLT powder and mixture of LLT powder and PS beads, implying that the bi-layered LLT with various thicknesses of dense and porous layers can be prepared by altering the amount of suspensions for filtration. No gap was observed at interface between dense and porous layers (fig. 2(a)). From this result, it can be said that Li ion can move from the porous layer to the dense layer smoothly and this bi-layered structure can be applied to a framework for the all-solid-state lithium battery. The porous layer has numerous ordered pores interconnected each other (fig. 2(c)). The pore size was about 1.8 μm, which was a smaller than diameter of PS beads template (3 μm) due to shrinkage of PS beads during calcination.

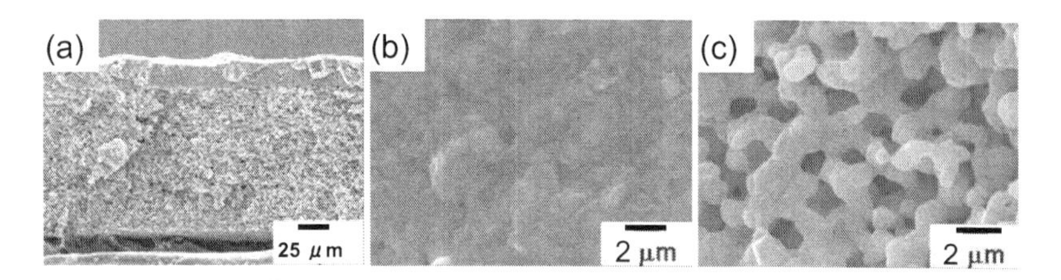

Figure 2. Cross sectional SEM image of bi-layered LLT (a) whole cross-section (b) magnified image of dense layer and (c) magnified image of porous layer

Figure 3 reveals XRD pattern of the bi-layered LLT. Most of diffraction peaks of the bi-layered LLT were assigned to LLT ($Li_{0.35}La_{0.55}TiO_3$). However, small peaks of impurity phase attributed to $Li_{0.33}La_{0.557}TiO_3$ were also observed. This impurity phase was formed by Li loss during calcination at high temperature as well as 1100 °C. This phase was also observed in honeycomb type LLT after calcination in our previous research[12].

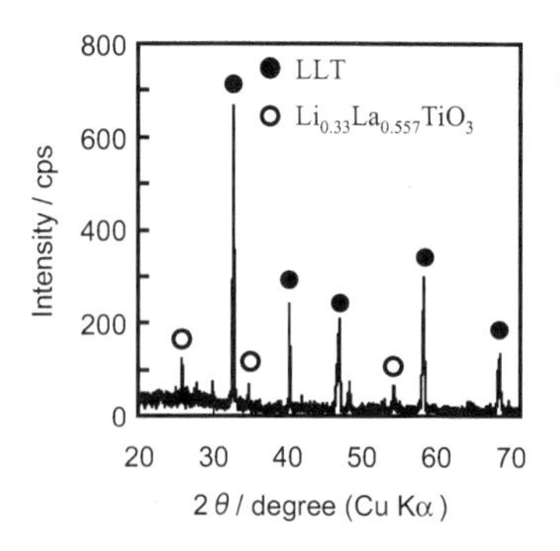

Figure 3. XRD pattern of bi-layered LLT

The AC impedance of the bi-layered LLT was measured. In complex impedance plot (fig. 4), a semicircle in the high frequency region and Warburg-type impedance in the low frequency region were observed. The intercepts of semicircle at high and low frequency sides were attributed to bulk and grain boundary resistances, respectively. The estimated conductivities of bulk and grain boundary were 8.0×10^{-4} and 2.6×10^{-5} S cm^{-1}, respectively. These values were comparable with reported values[13, 14].

Figure 5 shows cross-sectional SEM images of $LiMn_2O_4$ / bi-layered LLT composite. $LiMn_2O_4$ particles in the pores of LLT were clearly observed. However, as shown in a magnified image of porous layer near the dense layer (fig. 5(b)), some voids were still observed in bottom of the porous layer, while the porous layer near surface was filled better (fig. 5(c)). A volume fraction of $LiMn_2O_4$ referred to pore volume was about 40 %. The volume fraction was constant after the fourth time injection. In the case of 3DOM LLT membrane, the pores were filled by $LiMn_2O_4$ with nearly 80 % of volume fraction after the fourth time of the substitution process[7]. A reason of slightly lower fraction of bi-layered LLT is thought to be an existence of the dense layer. The sol can go into the pores from every direction of 3DOM LLT membrane. However, in the case of the bi-layered LLT, one side of the porous layer was closed by the dense layer.

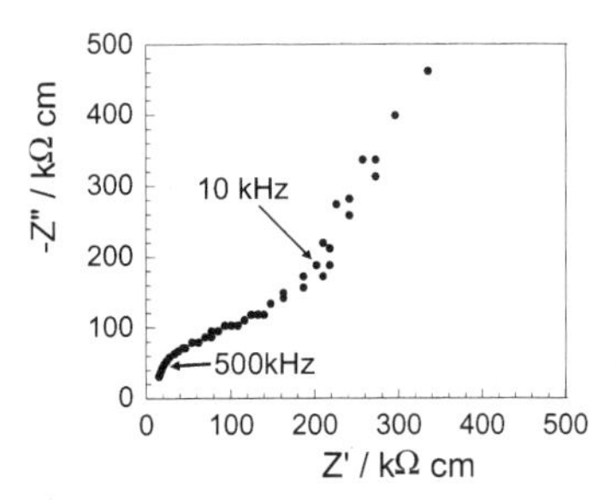

Figure 4. A complex impedance plot of bi-layered LLT.

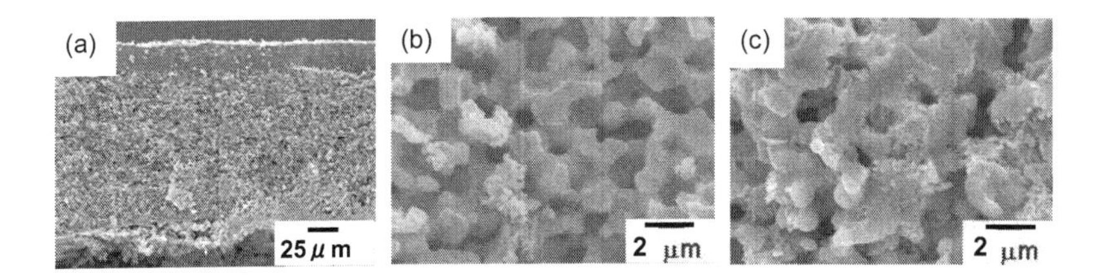

Figure 5. SEM images of $LiMn_2O_4$ / bi-layered LLT composite. (a) whole cross-section, (b) magnified image of porous layer near dense layer, and (c) magnified image of top of porous layer

Figure 6 exhibits charge and discharge curves of the bi-layered LLT with $LiMn_2O_4$ in its pores. Two potential plateaus were clearly observed at 4.0 and 4.1 V vs Li / Li^+ correspounding to redox of $Mn^{3+/4+}$ in $LiMn_2O_4$, revealing that $LiMn_2O_4$ in the pores worked as cathode. A large irreversible behavior was confirmed in the first charge / discharge cycle. This large irreversible capacity may be attributed to a degradation electronic path in the $LiMn_2O_4$ by volume expansion during charge process. The discharge capacity was 27 mA h g^{-1} which is 19 % of theoretical capacity (148 mA h g^{-1})[15]. Such low capacity was caused by low utilization of $LiMn_2O_4$ in the porous layer. Some voids were observed in the porous layer (fig. 5(b)), thus, some parts of $LiMn_2O_4$ were isolated electrically. This results in low utilization of $LiMn_2O_4$. Further study on optimizations of the solvent substitution method and structure of bi-layered LLT preparation are required.

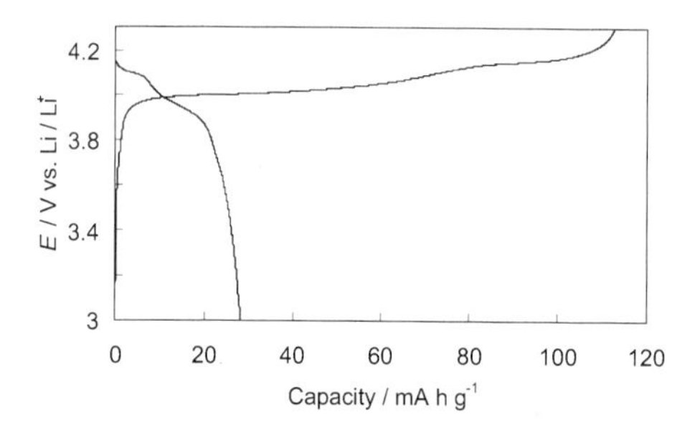

Figure 6. Charge / discharge curves of $LiMn_2O_4$ / bi-layered LLT composite measured at 0.05C

CONCLUSION

Bi-layered $Li_{0.35}La_{0.55}TiO_3$ (LLT) ceramics electrolyte which comprised of 3 dimensionally ordered macroporous (3DOM) layer and dense layer was fabricated by the suspension filtration method successfully. Thicknesses of the dense and the porous layers were about 23 and 105 μm, respectively. No gap was observed at interface between dense and porous layers. From this result, it can be seen that Li ion can move from the porous layer to the dense layer smoothly and it can be applied as a framework for the all-solid-state lithium battery. The porous layer was formed by uniform pores whose diameter was about 1.8 μm. After impregnation of precursor sol for $LiMn_2O_4$ into the pores following calcination, electrochemical property of $LiMn_2O_4$ / bi-layered LLT composite was also tested. The composite clearly showed charge and discharge behavior. This indicates that the composite can work as rechargeable battery. Discharge capacity of the composite was 27 mA h g^{-1}.

REFERENCES

[1]J. M. Tarascon, M. Armand, Issues and Challenges facing rechargeable lithium batteries, *Nature* **414**

359-367 (2001).

[2]K. Hasegawa, Y. Arakawa, Safety study of electrolyte solutions for lithium batteries by accelerating-rate calorimetry, *J. Power Sources* 43-44 523-529 (1993).

[3]J. R. Dahn, E. W. Fuller, M. Obrovac, U. Von Sacken, Thermal stability of Li_xCoO_2, Li_xNiO_2 and λ-MnO_2 and consequences for the safety of Li-ion cells, *Solid State Ionics* 69 265-270 (1994).

[4]K. Hoshina, K. Dokko, K. Kanamura, Investigation on Electrochemical Interface between $Li_4Ti_5O_{12}$ and $Li_{1+x}Al_xTi_{2-x}(PO_4)_3$ NASICON-Type Solid Electrolyte, *J. Electrochemical Soc.*, 152 A2138-A2142 (2005).

[5]K. Dokko, N. Akutagawa, Y. Isshiki, K. Hoshina, K. Kanamura, Preparation of three dimensionally ordered macroporous $Li_{0.35}La_{0.55}TiO_3$ by colloidal crystal templating process, *Solid State Ionics,* 176 2345-2348 (2005).

[6]H. Nakano, K. Dokko, M. Hara, Y. Isshiki, K. Kanamura, Three-dimensionally ordered composite electrode between $LiMn_2O_4$ and $Li_{1.5}Al_{0.5}Ti_{1.5}(PO_4)_3$, *Ionics* 14 173-177 (2008).

[7]M. Hara, H. Nakano, K. Dokko, S. Okuda, A. Kaeriyama, K. Kanamura, Fabrication of all solid-state lithium-ion batteries with three-dimensionally ordered composite electrode consisting of $Li_{0.35}La_{0.55}TiO_3$ and $LiMn_2O_4$, *J. Power Sources* **189** 485-489 (2008).

[8]V. Thangadurai, H. Kaacj, W. Weppner, Novel fast lithium ion conduction in garnet-type $Li_5La_3M_2O_{12}$ (M=Nb, Ta), *J. Am. Ceram. Soc.* 86 437-440 (2003).

[9]V. Thangadurai, S. Adams, W. Weppner, Crystal Structure Revision and Identification of Li^+-Ion Migration Pathways in the Garnet-like $Li_5La_3M_2O_{12}$ (M = Nb, Ta) Oxides, *Chem. Mater.*, 16 2998-3006 (2004).

[10]R. Murgan, V. Thangadurai, W. Weppner, Fast Lithium Ion Conduction in Garnet-Type $Li_7La_3Zr_2O_{12}$, *Angew. Int. Chem. Ed.* 46 7778- 7781 (2007).

[11]P. Knauth, Inorganic solid Li ion conductors: An overview, *Solid State Ionics*, 180 911-916 (2009).

[12]M. Kotobuki, Y. Suzuki, H. Munakata, K. Kanamura, Y. Sato, K. Yamamoto, T. Yoshida, Compatibility of $LiCoO_2$ and $LiMn_2O_4$ cathode materials for $Li_{0.55}La_{0.35}TiO_3$ electrolyte to fabricate all-solid-state lithium battery, *J. Power Sources*, **195** 5784-5788 (2010).

[13]Y. Inaguma, L. Chen, M. Itoh, T. Nakamura, Candidate compounds with perovskite structure for high lithium ionic conductivity, *Solid State Ionics* **70/71** 196-202 (1994).

[14]M. Itoh, Y. Inaguma, W. H. Jung, L. Chen, T. Nakamura, High lithium ion conductivity in the perovskite-type compounds $Ln_{12}Li_{12}TiO_3$(Ln=La,Pr,Nd,Sm) , *Solid State Ionics* **70/71** 203-207 (1994).

[15]R. K. Katiyara, R. Singhalb, K. Asmarb, R. Valentina, R.S. Katiyarb, High voltage spinel cathode materials for high energy density and high rate capability Li ion rechargeable batteries, *J. Power Sources* **194** 526-530 (2009).

AUTOGENIC REACTIONS FOR FABRICATING LITHIUM BATTERY ELECTRODE MATERIALS

Vilas G. Pol* and M. M. Thackeray

Electrochemical Energy Storage Department, Chemical Sciences and Engineering Division, Argonne National Laboratory, 9700 S. Cass Avenue, Argonne, IL 60439, USA

*To whom correspondence should be addressed. E-mail: pol@anl.gov

ABSTRACT

Autogenic reactions, based on the decomposition of chemical precursors at elevated temperatures under self generated pressures are being used to prepare a wide range of materials with interesting structural, morphological and technological properties. The technique is highly versatile; it can produce a wide range of carbon materials and architectures, carbon coated metal oxides and lithium metal phosphates, all of which are of interest as electrodes for lithium-ion battery applications. One particularly interesting example is spherical carbon, which is of interest a possible anode material. Carbon spheres can be prepared by the thermal decomposition of mesitylene (C_9H_{12}) or plastic waste in an enclosed reactor under nitrogen or argon at 700 °C. The solid carbon spheres, typically 2 – 4 μm in diameter, contain turbostratically disordered graphite sheets, the ordering of which can be increased by heating the spheres to 2400 °C under inert conditions. Heat treatment improves the electrochemical properties of these carbon electrodes; the initial electrochemical reaction of lithium occurs predominantly below 1 V, typical of a hard carbon electrode, generating a capacity of 307 mAh/g. The irreversible capacity loss on the first cycle (15.5 %) is significantly less than that observed for unheated samples, typically 60 %. On cycling at 51 mA/g (~C/5 rate) between 1.5 V and 5 mV, the carbon spheres provide a steady capacity of approximately 250 mAh/g.

INTRODUCTION

Autogenic reactions, which are distinct from solvothermal and hydrothermal processes, can be used to synthesize materials with interesting morphological properties; they have been discussed previously with respect to the fabrication of nanosized sulfide-, selenide-, boride-, phosphide-, nitride-, carbide-, and oxide compounds.[1] The nanostructures formed by this technique have been used for a variety of applications, such as hydrogen storage, catalyst supports, as well as for magnetic, optical and luminescent devices. In recent years, we have focused our attention on using autogenic reactions for fabricating both anode[2,-4] and cathode[5,6] electrode materials for lithium-ion batteries.

Graphite is an intercalation electrode which is the anode material of choice for today's lithium-ion batteries; it accommodates one lithium atom per C_6 (graphite) unit a few tens of millivolts above the potential of metallic lithium, thereby generating a theoretical electrochemical capacity of 372 mAh/g.[7] Hard carbons, which typically contain turbostratically-disordered graphitic sheets and that operate, on average, above the potential of lithiated graphite have been used to enhance the safety of carbon anodes.[8] Several synthetic methods, such as carbonization, high-voltage-arc electricity, laser ablation, or hydrothermal carbonization have been used to prepare various carbon materials with different degrees of crystallinity, size, shape, morphology and porosity.[9-12] For example, high-temperature hydrothermal carbonization produces olivary carbon particles[13] or carbon microspheres[14-16] from organic molecules. Meso-carbon micro beads (MCMB) with rounded edges, used by the lithium battery industry, are synthesized at extremely high temperatures under an inert atmosphere (2800 °C or higher).[17] Graphitic carbon-fibers, produced at 2850 °C from pitch-based carbon fibers have the advantage that they can minimize heat-generating side reactions during overcharge.[18]

We have recently demonstrated the ability of autogenic reactions to produce spherical carbon particles[19] and carbon nanotubes[20] using waste plastic bags as a feedstock. This research has highlighted this environmentally-attractive approach to synthesize novel or modified carbon-based materials. These 'upcycled' spherical carbon particles and carbon nanotubes have been evaluated as anode materials in lithium electrochemical cells[4]; they deliver a rechargeable capacity of more than 250mAh/g and cycle with >99% coulombic efficiency. In particular, the effect of heating the carbon spheres to 2400 °C for one hour on their morphological, structural and electrochemical behavior was discussed in reference 4. In this paper, we describe the autogenic synthesis of carbon spheres using mesitylene as the precursor, and the effect of heating the spheres at 2400 °C for a longer period of time (8 h) on their structural, morphological and electrochemical properties.

2. EXPERIMENTAL

The fabrication of spherical carbon particles was carried out by the thermal decomposition of 2g of mesitylene in an enclosed, but ventable, 10 ml Haynes 230 alloy autoclave reactor under argon. The reactor was heated to 700 °C with a heating rate of 20 °C/min; the reactor was held at 700 °C for 2h, which generated an autogenic pressure of approximately 500 psi. Thereafter, the reactor was cooled gradually to room temperature. The yield of carbon spheres was approximately 50 wt.%. The as-prepared carbon spheres (CS) were heated by Superior Graphite, Chicago, Illinois to 2400 °C for 8 h (CS-HT8) under inert conditions to enhance the graphitic character of the spheres.

Characterization

A high-resolution JEOL-7500 field emission scanning electron microscope (FE-SEM) was employed to analyze the morphology of the as-prepared CS and CS-HT8 samples. High-resolution SEM images of uncycled and cycled CS-HT8 electrodes were obtained to monitor the structural and morphological changes that had occurred during electrochemical cycling. Before the FE-SEM measurement, the cycled electrode was gently washed with dimethyl carbonate, (DMC, 99%) followed by an ethanol wash and dried under vacuum prior to collecting the scanning electron micrographs. Raman spectra of both uncycled and cycled CS-HT8 electrodes were obtained with an InVia Raman spectrometer (633 nm red laser with 10% intensity) at room temperature to monitor any changes in graphitic disorder that had occurred within the spheres as a function of electrochemical cycling.

Electrode fabrication

Electrochemical evaluations of CS-HT8 electrodes were undertaken using 2032 coin cells that contained a lithium metal foil counter electrode and an electrolyte consisting of 1.2M $LiPF_6$ in a 3:7 mixture of ethylene carbonate (EC) and ethylmethyl carbonate (EMC) solvents, sourced from Tomiyama, Japan. Cells were assembled in a helium-filled glove box. The working electrode consisted of 85 wt% of CS-HT8, 8 wt% carbon black, and 7 wt% polyvinylidene difluoride binder laminated onto a copper foil current collector. Cells were galvanostatically charged and discharged between 1.5 V and 5 mV at 51 mA/g (~5C rate) at room temperature.

RESULTS AND DISCUSSION

Figures 1a-b shows SEM images of the as-prepared carbon spheres at various magnifications, depicting 2 – 4 μm sized particles with smooth surfaces. The particles are fairly monodispersed, almost perfectly spherical in shape and have a solid nature.[21] Corresponding images of spheres that had been heated to 2400 °C for 8 h in an inert atmosphere are shown in Figures 1c-d. It is apparent from Figures 1c-d that the heat treatment process had a negligible effect on the spherical shape and overall morphology of the particles, confirming the remarkable stability of the spheres when heated to an extremely high temperature. EDX elemental analyses confirmed that the spheres were essentially carbon; no impurities could be detected by this method in both as-prepared CS and heated CS-HT8 samples. Powder X-ray diffraction patterns and Raman spectra of CS and CS-HT8 products were analogous to those obtained when polyethylene plastic waste was used as the precursor for the autogenic reaction.[4]

Figure 1: Scanning electron micrographs of (a-b) as-prepared CS and, (c-d) CS-HT8 with different magnification.

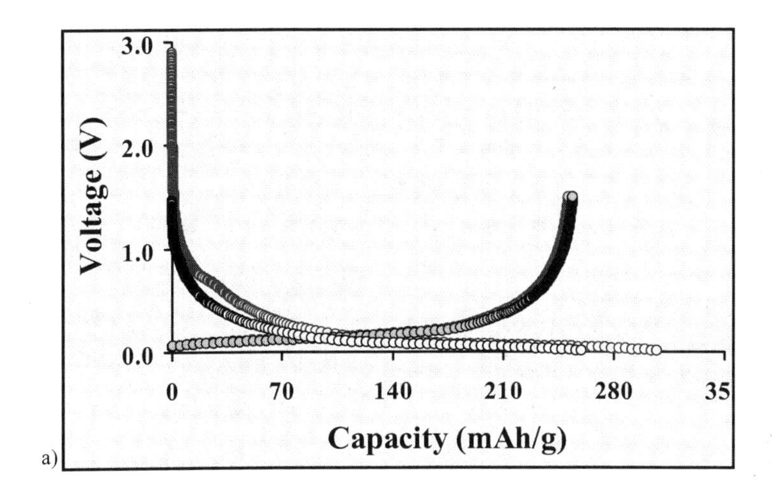

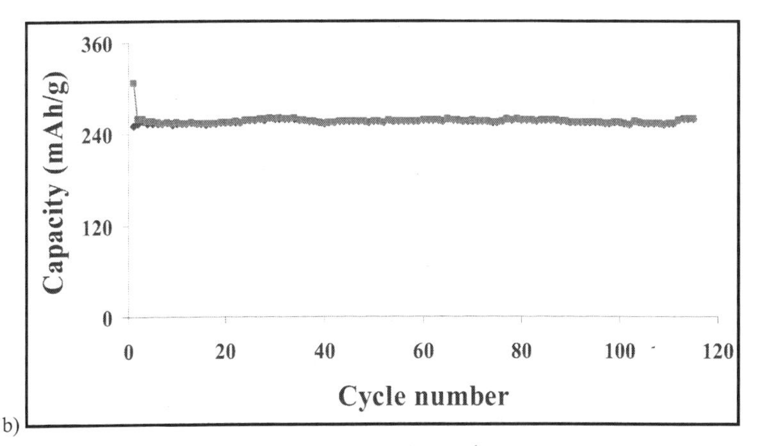

Figure 2: a) Discharge-charge profiles for the 1st and 2nd cycles, and b) capacity vs. cycle number of a Li/CS-HT8 cell at 51 mA/g (~5C) rate; 1.5 V – 5 mV.

The 1st and 2nd discharge/charge cycles of a Li/ CS-HT8 cell, cycled at a constant current density of 51 mA/g (~5C rate) between 1.5 V and 5 mV are shown in Figure 2a. The initial reaction of the carbon spheres with lithium, which occurs predominantly below 1 V, generated an electrochemical capacity of ~307mAh/g carbon, whereas the subsequent charge and discharge yielded 251 and 259 mAh/g, respectively. The 15.5% irreversible capacity loss between the first and second discharge reactions is attributed to the well known phenomena of electrolyte reduction and solid-electrolyte interphase (SEI) formation.[22] Our previous experiments on spherical carbon electrode particles that had been heated at 2400 °C for 1 hour showed a 20% irreversible capacity loss between the first and second discharge.[4] We tentatively attribute this improvement in electrochemical behavior to a slightly reduced surface area and to an increase in the graphitic order within the carbon spheres as a result of the longer heat-treatment time. The sloping charge/discharge profile is typical for a hard carbon containing turbostratically-disordered graphitic sheets[23] on both sides of which lithium can bond to carbon during the electrochemical charge of a lithium-ion cell.

After the initial charge/discharge cycle, the electrochemical capacity of the CS-HT8 electrode particles remains unaffected on cycling, as shown in a capacity vs. cycle number plot of a Li/CS-HT8 cell in Figure 2b. Continued cycling is remarkably stable for a lithium half cell; after 120 cycles, the discharge and charge capacities for the carbon electrode were 258 and 255 mAh/g, respectively, reflecting a columbic efficiency of more than 98%. We are currently attempting to improve the cycling capacity by increasing the graphitic content even further by annealing the carbon spheres at 2800 °C under inert conditions.

Raman spectroscopy was employed to obtain a better understanding of the structural features within the CS-HT8 spheres and to complement the FE-SEM data that was used to monitor the morphological changes during lithiation and delithiation in uncycled and cycled electrodes. The Raman spectra of CS-HT8 electrode particles before (Figure 3, bottom), and after (Figure 3, top), 100 discharge-charge cycles confirm the noticeable change to the disordered and graphitic carbon components. The most prominent features in the Raman spectra of uncycled CS-HT8 electrode are the G (graphitic) band appearing at 1585 cm^{-1} and the second-order G' band at about 2665 cm^{-1}.[24] The G and G' bands in the Raman spectrum are an indication of sp^2 carbon networks.[25] The D (disordered) band at about 1330 cm^{-1} and the D'- band at about 1620 cm^{-1}, are generally attributed to defect induced Raman features usually observed in poorly ordered graphite.[24] The integrated intensity ratio, ID/IG, for the D band and G band is widely used for characterizing the defect quantity in graphitic materials.[24] In the uncycled CS-HT8 electrode, the ID/IG ratio is 0.61, while in the cycled CS-HT8 electrode this ratio is 1.25, indicating that the graphitic disorder increases in the carbon sphere electrode during electrochemical cycling. We attribute this phenomenon to local strain induced within the CS-HT8 particles during the repeated charging (lithiation) and discharging (delithiation) reactions.

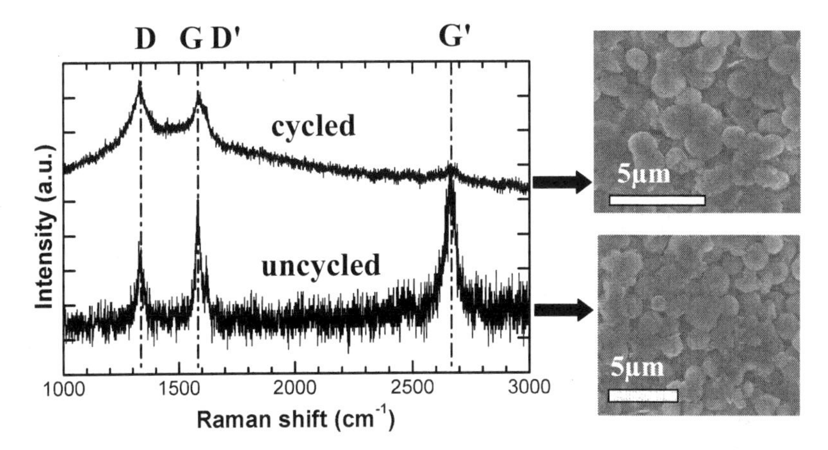

Figure 3: Raman Spectra of as-prepared CS-HT8 electrodes (bottom) and corresponding SEM image on right, and cycled CS-HT8 electrode (top), after 100 charge-discharge cycles and corresponding SEM image on right.

The corresponding SEM images of the as-prepared carbon sphere electrode and after being cycled for 100 times are almost analogous. Note that in the images, the PVDF binder and the conducting carbon black additive surround the electrode particles, providing a 'glue' and enhancing electrical contact between particles, respectively. The most remarkable feature, however, is that the spherical shape and general morphology of the SCP-HT8 electrode particles were retained on cycling, unlike as-prepared carbon spheres that are not subjected to a heat-treatment step, and which tend to disintegrate on electrochemical cycling.[4] Because the first cycle irreversible capacity loss associated with Li/CS-HT8 electrodes (8 hour heat-treatment) is lower than that provided by CS-HT1 electrodes (1 hour heat-treatment) and because cells containing the former electrodes cycle with slightly superior coulombic efficiency, the data seem to suggest that a longer heat-treatment time results in a lowering of the surface area, reducing contact with electrolyte and the extent of SEI formation. Finally, it appears that the heat-treatment step emphasizes the possibility of tailoring the relative amounts of disordered and graphitic carbon in the spheres to fine tune the average operating potential of carbon sphere electrodes.

CONCLUSIONS

Autogenically fabricated carbon spheres do not change their shape or size after annealing at 2400 °C for 8 h under inert conditions and during electrochemical cycling with lithium. When used as electrodes in lithium half cells, these materials deliver a stable capacity of approximately 250 mAh/g between 1.5 V and 5 mV for hundreds of cycles, despite an apparent decrease in the graphitic character of the carbon. The superior electrochemical properties of heated carbon spheres, which display only 15.5 % irreversible capacity loss, are attributed to an annealing process during the heat-treatment step that lowers the surface area and increases graphitic order within the spheres. Efforts are underway to increase the graphitic character of carbon sphere electrodes even further by heating them to higher temperatures, e.g., 2800-3000 °C in an inert atmosphere, with the hope of increasing their ability to generate a higher capacity, closer to that of graphite.

ACKNOWLEDGEMENTS

This work was supported by the Center for Electrical Energy Storage: Tailored Interfaces, an Energy Frontier Research Center funded by the U.S. Department of Energy, Office of Science, Office of Basic Energy Sciences. Use of the FESEM and Raman spectroscopy facilities at Argonne's Center for Nanoscale Materials (CNM) is acknowledged. Jorge Ayala and Francois Henry of Superior Graphite, Chicago, Illinois are thanked for undertaking the heat-treatment of the carbon spheres.

REFERENCES

[1]V. G. Pol, S. V. Pol, and A. Gedanken, Novel Nanostructures of Borides, Nitrides, Phosphides, Carbides, Sulphides, Selenides and Oxides Fabricated by Dry Autoclaving of the Precursors at Elevated Temperature, *Adv. Mater. in press* (adma.201001490).

[2]V. G. Pol, M. M. Thackeray, S. H. Kang, C. S. Johnson, and J. M. Calderon Moreno, Autogenic Reactions for Preparing Carbon-encapsulated Nanoparticulate TiO_2 Electrodes for Lithium-ion Batteries, *J. Power Sources,* **195,** 5039-43 (2010).

[3]S.-H. Kang, V. G. Pol, I. Belharouak, and M. M. Thackeray, Comparison of High Capacity xLi_2MnO_3. 1-x $LiMO_2$ (M=Ni,Co,Mn) Cathodes in Lithium-Ion Cells with $Li_4Ti_5O_{12}$- and Carbon-Encapsulated Anatase TiO_2 Anodes, *J. Electrochem.Soc.* **157** , A267-71 (2010).

[4]V. G. Pol and M. M. Thackeray, Spherical Carbon Particles and Carbon Nanotubes Prepared by Autogenic Reactions: Evaluation as Anodes in Lithium Electrochemical Cells, *Energy & Environ. Sci.,* in press, DOI:10.1039/C0EE00256A.

[5]A. Odani, V. G. Pol, S. V. Pol, M. Koltypin, A. Gedanken and D. Aurbach, Testing Carbon-coated VOx Prepared via Reaction under Autogenic Pressure at Elevated Temperature as Li-insertion Materials, *Adv. Mater.* **18**, 1431-37 (2006).

[6]M. Koltypin, V. G. Pol, A. Gedanken and D. Aurbach, The Study of Carbon-coated V_2O_5 Nanoparticles as a Potential Cathodic Material for Li Rechargeable Batteries, *J. Electrochem. Soc.* **154**, A605-13, (2007).

[7]T. Ohzuku, K. Iwakoshi, and K. Sawai, Formation of Lithium-Graphite Intercalation Compounds in Nonaqueous Electrolytes and Their Application as a Negative Electrode for a Lithium Ion (shuttlecock) Cell, *J. Electrochem. Soc.*, **140**, 2490-98 (1993).

[8]E. Buiel and J.R. Dahn, Li-insertion in Hard Carbon Anode Materials for Li-ion Batteries, *Electrochimica Acta*, **45**, 121-125 (1999).

[9]Q.L Lin, H. Y. Tang, D.Y.Guo, and M.Z. Zheng, Preparation and Properties of Carbon Microbeads by Pyrolysis of N-phenyl Maleimide Modified Novolac Resin, *J. Anl. and Appl. Pyrolysis,* **87,** 276-81, (2010).

[10]B. C. Satishkumar, A. Govindaraj, C. N. R. Rao, Bundles of Aligned Carbon Nanotubes Obtained by the Pyrolysis of Ferrocene-Hydrocarbon Mixtures: Role of the Metal Nanoparticles Produced In situ *Chem. Phys. Lett.* **307**, 158-162 (1999).

[11]P. W. Barone, S. Baik, D. A. Heller, and M. S. Strano, Near-infrared Optical Sensors Based on Single-Walled Carbon Nanotubes, *Nat. Mater.* 4, 86-U16 (2005).

[12]H. Dumortier, S. Lacotte, G. Pastorin, R. Marega, W. Wu, D. Bonifazi, J.-P. Briand, M. Prato, S. Muller, A. Bianco, Functionalized Carbon Nanotubes are Non-cytotoxic and Preserve the Functionality of Primary Immune Cells, *Nano Lett.* **6**, 1522-28 (2006).

[13]T. Luo, L. S. Gao, J. W. Liu, L. Y. Chen, J. M. Shen, L. C. Wang, and Y. T. Qian, Olivary Particles: Unique Carbon Microstructure Synthesized by Catalytic Pyrolysis of Acetone, *J. Phys. Chem. B* **109**, 15272-77 (2005).

[14]Y. Z. Mi, W. B. Hu, Y. M. Dan, and Y. L. Liu, Synthesis of Carbon Micro-spheres by a Glucose Hydrothermal Method, *Mater. Lett.* **62**, 1194-96 (2008).

[15]K. C. Park, H. Tomiyasu, S. Morimoto, K. Takeuchi, Y. J. Kim, and M. Endo, Carbon Formation Promoted by Hydrogen Peroxide in Supercritical Water, *Carbon* **46**, 1804-08 (2008).

[16]X. C. Ma, F. Xu, Y. Du, L. Y. Chen, and Z. D. Zhang, *Carbon,* Copper Substrate-assisted Growth of Ellipsoidal Carbon Microparticles, **44,** 179-81. (2006).

[17] J. Yao, G.X. Wang, J.-H. Ahn, H.K. Liu, S.X. Dou, Electrochemical Studies of Graphitized Mesocarbon Microbeads as an Anode in Lithium-ion Cells *J. Power Sources*, **114,** 292-297 (2003).

[18]S. Hossain, Y. K. Kim, Y. Saleh, and R. Loutfy, Overcharge Studies of Carbon Fiber Composite-based Lithium-ion Cells, *J. Power Sources,* **161**, 640-647 (2006).

[19]V. G. Pol, Upcycling: Converting Waste Plastics into Paramagnetic, Conducting, Solid, Pure Carbon Microspheres, *Environ. Sci. & Tech.*, **44 (12)**, 4753-59 (2010).

[20]V. G. Pol and P. Thiyagarajan, Remediating Plastic Waste into Carbon Nanotubes, *J. Environ. Monitoring,* **12**, 455-59 (2010)

21V. G. Pol, M. Motiei, A. Gedanken, J. Calderon-Moreno and M. Yoshimura, Carbon Spherules: Synthesis, Properties and Mechanistic Elucidation, Carbon, 42, 111-16 (2004).

[22]V. Eshkenazi, E.Peled, and Y.Rosenberg, Study of Lithium Insertion in Hard Carbon Made from Cotton Wool, *J. Power Sources,* **76**, 153-58 (1998).

[23]J. R. Dahn, T. Zheng, Y. Liu, and J. S. Xue, Mechanisms for Lithium Insertion in Carbonaceous Materials, *Science,* **270**, 590-93 (1995),

[24]F. Tuinstra and J. L. Koenig, Raman Spectrum of Graphite, *J. Chem. Phys..* **53**, 1126-30 (1970)

[25]M. A. Pimenta, G. Dresselhaus, M. S. Dresselhaus, L. G. Canc,ado, A. Jorioa and R. Saitoe Studying Disorder in Graphite-Based Systems by Raman Spectroscopy, *Phys. Chem. Chem. Phys.,* **9**, 1276–91 (2007).

Author Index